Advanced Technology
for Smart Buildings

For a complete listing of titles in the
Artech House Power Engineering Library,
turn to the back of this book.

Advanced Technology for Smart Buildings

James Sinopoli

ARTECH
HOUSE

BOSTON | LONDON
artechhouse.com

Library of Congress Cataloging-in-Publication Data

A catalog record for this book is available from the U.S. Library of Congress.

British Library Cataloguing in Publication Data

A catalogue record for this book is available from the British Library.

Cover design by John Gomes

ISBN 13: 978-1-60807-865-3

© 2016 ARTECH HOUSE
685 Canton Street
Norwood, MA 02062

10 9 8 7 6 5 4 3 2 1

Contents

Introduction

This book deals with the deployment of advanced technology in buildings. It's meant to be informative and useful. It covers major control systems and attributes of a smart building. Smart buildings are distinguished by the use of advanced technology. Advanced technology can come in many forms and can be deployed in many ways. Some of the advanced technologies addressed in the book are proven to provide value and results, while others are new to the marketplace and appear very promising. The conclusion is that advanced technology can improve building performance, provide superior operation of building systems, lower the cost of operations, reduce service orders and maintenance, deliver higher satisfaction to tenants and occupants, and generate greater financial returns for the building owner.

A smart building is consistent with the holistic principal that it's the whole is greater than the sum of the parts. It's not one system or attribute that makes a building smart, it's a combination of systems and practices that comprise a total approach. This book provides a summary with details regarding the roles of different participants in the smart building technology implementation process, various control systems, practices such as enterprise data management or energy management as well as emergent technology like indoor positioning systems, eye tracking systems, and software analytic applications.

The primary focus in the application of advanced technology is ongoing building operations. It's estimated that 75–80% of a building's lifecycle costs and duration is related to building operations. Whether it's an existing building or new construction the building owner who is not turning over real estate quickly has to take a long term view of the building operations. Astute owners will make longer term investments in smart buildings systems, equipment and building attributes if they're to be assured of positive results, value, and return on investment.

There has been constant innovation in building construction since the first building was built for purposes other than just personal shelter. In the 19th and 20th centuries major innovations such as running water, sewage systems, concrete, construction cranes, power equipment, electricity, and advanced materials all had an enormous impact on buildings. Likewise modern elevators facilitated high rise buildings and denser urban cities. Today's technology produces new materials and innovations as well. The impetus for these advancements is related to the global focus on energy and sustainability with a growing and dynamic marketplace of new energy products, services, and companies. Efficient geothermal heat pumps, coatings for windows, companies aggregating utility demand response, microgrids, direct current electrical infrastructure, and renewables, are some examples.

The other major driver of advanced technology in buildings is *information and communications technology* (ICT). In many ways ICT is much broader than energy and sustainability because billions on the planet are now habituated to ICT via smartphones, tablets, and computers in their daily routine. The eventual vision to be created from ubiquitous ICT devices is the *Internet of Things* where any sensors, devices, or software applications that can be connected will be connected. These networks will be things to things, people to things, and people to people. ICT is a major underlying foundation of smart buildings, as well as smart cities.

The penetration of ICT into building control systems has started with typical building control system communications protocols such as BACnet, Lonworks, and Modbus already incorporating the transport of those protocols over an IT network. Many traditional building systems such as video surveillance, access control, and other components now operate over an IT network. We also have the crossover of IT cabling standards for building automation systems, automation controllers for mechanical equipment using Ethernet or Wi-Fi, and power to devices provided by Power-over-Ethernet (POE), an IT creation. In addition, the building management systems are IT devices; computers and servers, operating systems, databases, software applications, and IP addresses.

The facility management industry is in its infancy using IT as part of managing a building. However, facility management has already seen value in advanced building management tools, enterprise data management, data analytics, software applications, and leveraging and incorporating ICT. With the initial positive results of incorporating ICT into building construction, we can expect increased utilization. The penetration of IT has created issues organizationally; that is, how IT and Facility Management departments work together in their responsibilities to monitor and maintain building systems. Eventually new organizational structures will be created, such as an umbrella department for IT and facility management, or embedding IT experts in facility management.

Despite recent advancements we're not close to the full potential of automation and advanced technology in improving and optimizing building performance. Increased automation would not only improve a buildings' performance, but, will also support the facility management challenge to manage increasingly complex buildings at a time when the skills require-dare constantly changing and in short supply. An example of where we are currently and where we need to go would be a software application like fault detection and diagnostics for HVAC systems, probably the most effective building analytic software application today, (still only half a loaf.) What if we had an application that could not only automatically detect system faults but also automatically repair the faults (perhaps something similar to an autopilot?)?

The autopilot for airplanes was born one hundred years ago. We've now developed driverless automobiles. Why can't a building operate via auto pilot? Not only can buildings have autopilots but they should. Buildings are not airplanes or driverless cars, but, the idea of total automation should propel the control system and automation industry forward.

The roadmap to advanced, automated, and smart buildings involves several key issues the industry must address. One is the use of granular data. Granular data provides for more precision in properly managing specific spaces within a building, which can result in squeezing out the smallest amount of excess energy consumption and increase occupant satisfaction. Going granular will mean more sensors, tailored controls for individual spaces and a bit more investment which would ultimately lead to improved and less costly building operations. Another issue is policies and logic. For a building to be fully automated it requires the logic or policies of the automation use an array of data, data sources, and predetermined rules. As buildings become more complicated the decisions on their performance become more complex as variables increase. The third issue is data mining and the use of data analytics applications. Analytics for the HVAC system has provided outstanding outcomes. We need to applythat template to other building systems. Such applications are based on rules of how the system should optimally operate, generally obtained from the original design documents, and monitoring key data points in real-time. For those systems that are not process based, applying analytics generally uses statistical monitoring of key performance indicators (KPIs) to monitor outliers. Accurately analyzing data in a building with enhanced levels of automation is crucial as that data will be the foundation of the logic and policies of the automation process. The remaining issue is the increased use and maintainenance of sensors. Yes, highly automated buildings will need additional sensors and metering. With the market moving increasingly into developing information based on data, we need to generate more data.

Smart buildings encompass many things, but, the primary goal is the use of building technology systems to enable enhanced services and the efficient operation of a building for the betterment of its occupants and building management. The main drivers of smart buildings are the positive financial impacts of integrated systems, energy conservation, greater systems functionality, and the continuing evolution of technology.

Contents

The Role of Owners and Architects in a Smart Building

One of the tried and true adages in the design and construction industry is that "there are no good projects without good owners". The building owner is in the driver's seat when it comes to the necessities in a newly constructed or renovated building. They have devised the project and will fund and pay for design and construction. As the eventual purchaser of the building, the owner obviously has significant input and a tremendous stake in a successful project.

A good owner is a leader who can clearly communicate, is transparent in his decision-making, listens to experts, and encourages collaboration and innovation among the designers and contractors. Typical considerations for the owner involve the purpose of the building or space, budget, time constraints, the complexity of design, physical conditions, economic conditions, project sequencing, the project delivery method, as well as legal restrictions and environmental impacts.

Prior to committing to a project, building owners will commission preliminary studies to establish the project's feasibility. The building owner (especially for a commercial building) wants to know if the idea is economically sound. Some of the variables that may be assessed are general economic conditions, specific situations of the area or community where the building is to be located, projected population growth, land prices, and current cost of construction. If the project is a commercial building or development, some of the initial studies will also be related to similar existing competing businesses as well as an overall assessment of the business potential and climate.

The owner's team usually consists of the owner (or a representative) a construction manager, possibly an existing facility manager or engineer, and major tenants or the overall facility user. The owner will select an architect who will then assemble the design team. Two other project teams may be constituted as well. One would be the contractor team which includes the contractor or design-builder, the contractor's Project Manager, Construction Manager, Superintendent and any subcontractors. A supplier team will also be established, involving representatives from equipment or material manufacturers, independent product representatives and suppliers or distributors.

There are a number of approaches to the overall delivery of a building project. Probably the most used and somewhat traditional method is design, bid, and build. In this process, the design of the building is completed, then bid out to qualified contractors; with the competitive bidding used to determine the lowest cost bidder. Government funded projects generally use this method. The downside here is that lowest cost doesn't necessarily mean the best value for the owner.

A second approach is design, negotiate, and build. This is a more informal process where the contractor is involved in developing the costs and negotiating a contract to construct the project based on some stage of the design. In this case oftentimes the owner is looking for specific expertise, wants a notable architect or needs to expedite the schedule.

Another approach is design-build. In the design-build delivery method the owner signs on with a single entity for the complete design and construction of a project, providing design and construction under a single contract with the owner. Generally the design and the construction companies enter into a joint venture or one entity subcontracts to the other. At some point after initial design the sole entity provides a guaranteed maximum price for construction. The potential benefits of this approach are greater collaboration between designers and the contactors with fewer change orders or variances (since the designers and contractor are working as one team) and better adherence to the project schedule. This approach may save

Figure 1.1   ©Issola, used under CC BY-SA 3.0.

time and money, as well as reduce the owner's risk and potential litigation. Researchers at Pennsylvania State University found that design–build projects are generally delivered faster than those projects constructed under the design -bid-build process.

Often a building owner will hire a construction management company to manage the project from conception to completion in order to supplement the owner's staff and role. A construction management company will either advise the owner or act as a contractor. When acting as a contractor, they are referred to as construction management at-risk. Typically the building owner will bring in the construction management company prior to completion of the design and then incorporate the architect and construction management company into one entity or contract. Once contracted the construction management and the architect design teams review and evaluate the project, eventually coming up with a guaranteed maximum price. There are architectural and engineering firms that can perform construction management at risk as well. The benefits to this approach are enhanced coordination with the design team and contractors, better cohesion in the project team and reduced risk.

These different approaches to project delivery are vitally important because they define roles, legal responsibilities and risks, and profoundly impact the schedule, costs, and quality of the building.

1.1 Design Teams

The lead architect is typically the main interface with the building owner. It's the architect that develops the owner's facility program and assembles a design team, both of which are critical to the overall success of the project. With such a prominent role the architect heavily influences just how smart the building will be. The design team's basic responsibility is to transform the owner's vision and facility program into a detailed design. The team determines the design requirements, specifies and draws up the project, produces the construction documents and eventually administers the construction contract.

For a smart building it's critical to select design team members that are open minded, innovative, technologically savvy, and experienced. If out of habit engineers and designers just select old legacy designs they simply short change the owner. Due to the influence of energy and sustainability concerns, as well as the use and penetration of IT into building systems most designers have had some experience with advanced technology and understand the concept of smart buildings. Architects should obviously also be experienced dealing with buildings becoming increasingly complex; resulting in additional building systems, potential operational challenges for the owner, and design teams growing as representatives of various. For example, if you're trying to deploy solar panels, wind turbines, or a water reclamation system you're likely to bring in several different specialists to represent the latest in technology and design.

1.2 Facility Programming

The initial phase of a project is Project Conception or Facility Programming and is sometimes referred to as visioning meetings to lay out how the owner's requirements and objectives can be transformed into a clear or detailed concept or model. Such sessions may address current pain points, review emerging technology and innovations, and evaluate what other building owners have experienced.

The development of a facility program will be led by the architect in collaboration with specialized facility programmers, engineers, consultants, facility managers, contractors and manufacturers. It's a creative iterative process which teases out the owner's real objectives, values, and preferences, and identifies the needs and considerations related to aesthetics, economics, regulatory issues, energy, sustainability, and functionality. The result is the owner's unique facility plan which forms the foundation of the design and construction. (For more on programming see the classic book "Problem Seeking" by HOK, an architectural programming primer).

During this early programming activity the discussion of a smart building, automation, advanced technology, system integration, emerging technology, tenant amenities, building management systems, and analytic software applications takes place. Facilitating such a discussion requires participants able to talk about emerging technologies, results of such deployments at other projects, as well as the financial and operational benefits. Many times site visits to projects that include smart building attributes are educational for owners and their design team. The goal of the project conception meeting should be an openness to innovation and a focus on the facility lifecycle and building performance.

Some potential obstacles during project conception include designers or participants that continue to provide legacy designed systems; designers that have a specification for a system and keep repeatedly using an aged specification. While part of the motivation of designers may simply be their familiarity and positive experience with specific equipment or manufacturers, there is an incentive to reuse specifications and merely change the client name because it takes much less time.

Without focus on the potential use of advanced technology, options for system integration, and consideration of emerging technology and successful deployment of smart building goals in other buildings, it will not become an integral part of the building program. When the idea of a smart building becomes an afterthought, possibly identified later in the design process, its consideration may be disruptive and potential benefits diminished because of existing design decisions and concerns.

There are prominent features of the facility program that will be mandated by government regulations. There will be attributes that will be influenced by third-parties, the most obvious being a green certification program. Assuming the owner is seeking such certification, the guidelines provide necessary benchmarks and parameters for the design, construction, and operation of buildings likely to be in the facility program. Similar to green certification is the Smart Buildings Institute (SBI) and other organizations' certifications which focus on advanced technology, system and data integration and building operations, thus providing a framework and detailed properties for a smart building.

So one role for architects, perhaps their most important role in a smart building, is simply putting advanced technology and smart building operations on the agenda, then explaining the technology and economics to the owner and incorporating the main tenets of this approach in the overall facility program.

Surely architects understand that the control, monitoring, and building automation systems are essential components in the smart building. These systems are the dynamic features or facets of the building; the nervous system allowing for timely or real-time adjustments in the building's environment as well as optimal operational performance related to life safety, comfort, security, energy and an overall healthy atmosphere.

Architects must also understand that it's not just control systems that make up a smart building. The fixed attributes of the building such as the initial siting, the structure, envelope, windows, interior layout, materials, etc. also play a major role in how smart the building is and how well the building will operate. The best building control systems cannot compensate for the worst building construction and layout; in the same way, the best structure cannot overcome the worst building control systems. Both are critical in creating a smart and well-designed building. What follows are some of the functions and responsibilities of the architect and how they play a crucial role in designing, constructing, and operating a smart building.

1.3   Siting the Building

Architects frequently help the building owner with the process of selecting and acquiring the building's site for new construction, or in existing buildings assessing current conditions and updating a survey. Why is the site selection process so important to being a smart building? Because it is a long term, 40 to 100 year decision. The specifics of a site, the topography, climate, and available public utilities will affect the design and construction of the building, possibly incorporating specialized building systems such as seismic, tilt, corrosion, and ground pressure monitoring. Also, the general area surrounding the specific site is critical; proximity to transportation infrastructure, to other businesses, schools, and skilled labor pools may be important to the long term success of the building.

1.4   Materials

The architect and the design team will select the materials used in the building. These decisions are often a balancing act between constructability, aesthetics, durability, regulations and affordability. These considerations are important because materials deteriorate and wear out. The unpleasant result may be condensation, corrosion, stains, moisture retention, bending, rot, fungus, and a host of other negative properties. Materials are critical because they will affect the ongoing cost and ease of maintenance and operations, requiring servicing, cleaning, repairing or replacing. The selection of materials should focus on the long term use factoring in the cost of the materials as well as the minimization of maintenance.

Today's building materials must also be sustainable; that is, materials obtained from renewable, recycled, or replaceable sources. Products must be analyzed on a lifecycle basis, taking into account the environmental impacts of extracting the resources for the product, the manufacturing process, the materials transport and considering any impacts on material installers or building occupants. Guidelines for selection of materials are available from USGBC, the AIA's environmental resource guide and the environmental lifecycle assessment specified in ISO 1400 standards

1.5   Coordination

Architects are generally tasked with coordinating the information and work of the design team members. Later in the construction phase a construction manager may coordinate the work of the contractors. Coordination of design and construction is especially essential when building control systems are being integrated, a very critical element of a smart building. It's not enough to simply state in a specification that system A has to be integrated to System B; such pronouncements are too vague, don't identify and detail the responsibility of each party and oftentimes can result in finger-pointing and delays. As the design team leader, the architect should endorse a compliance statement for system integration responsibilities that each of the designers and engineers can acknowledge and insert in their particular specification that would cover the designers as well as the contractors. The statement may address standards for communications protocols, data format, submission of drawings and data points, responsibilities for any hardware, timing of the integration, the contractors' role in commissioning, etc.

1.6   The Handoff to Operations

As leader of the design team, the architect along with the contractors have responsibilities for the handoff from construction to operations. The design specifications must address important elements of the transition: startup procedures, closeout requirements, operation and maintenance data, preventative maintenance instructions, and facility operation procedures. A poor transition can mean the building launch and operations get off to a bad start, never fully recover or only catch-up after considerable cost, effort, and angst.

Studies have identified significant inefficiencies in the capital facilities marketplace primarily related to the lack of exchange of data. The inefficiencies were estimated to be in the billions of dollars. More interesting was the fact that two-thirds of that cost are borne by owners and operators, primarily during ongoing facility operation and maintenance. Other studies have been done regarding the use of Building Information Modeling (BIM), a tool for design, construction, and exchange of data primarily used in design and construction fabrication but very little in building operations. The point is that the handoff from construction completion to operations is often inadequate, setting up suboptimal operations from the start. Better handoffs prescribe that we embed operations and maintenance into every aspect of design and construction.

Contents

Measuring the Performance of a Building

With increased global awareness of energy consumption and sustainability we tend to measure the performance of a building based on its energy consumption, comparing its energy consumption against some base minimum energy performance such as the ASHRA E/IESNA Standard 90.1. The larger the improvement over the base energy performance the higher the building is performing. If a new building has exceeded the estimated base performance by 42% or an existing building by 35% (LEED criteria for optimizing energy performance) they have attained the highest level of building performance. But, what if this existing building has exceeded the base energy performance by 35% but is unsafe, unprofitable, or is not satisfying it occupants? Is it still a high performance building?

Energy consumption and sustainability are critical in buildings, but, the view that only energy defines a building's performance is myopic.

The fact is that most buildings are too complex to be evaluated on energy consumption alone. If you were shopping for an automobile would you base your buying decision solely on miles-per-gallon or kilometers-per-liter of gasoline? If so we'd all be driving single-seat cars with 4-cycle engines. Building performance needs to be defined more broadly, more holistically. Aside from energy and sustainability we need to examine other factors of a building's performance. While energy and sustainability are important unto themselves they also influence or affect other building performance factors, especially the financial aspect; that is, a large part of the motivation behind conservation, alternative energy sources, demand response, and so forth is to save money.

2.1 Financial Metrics

All buildings from modest houses to the tallest skyscrapers have financial aspects. Building is a business with defined industry sectors dealing with design, construction, management, and investments. Financial concerns cut across the lifecycle of a building; the construction or acquisition cost of the building, operating costs and the asset valuation of the building.

The financial metrics of a building can be almost endless, however there is published data for comparable commercial office or investment buildings a building owner can use to evaluate whether their building is above or below average financial metrics of similar buildings. For example, the Building Owners and Managers Association (BOMA) annually publishes the "Experience Exchange Report" which is based on data collected from thousands of buildings. The report allows users to examine expenses and income for similar type buildings in a common locale or submarket. The examination of basic balance and income reports for a building for profit and loss, increases or decreases to income, expenses and asset valuation can be used to judge financial performance.

Perhaps the major financial metric of a building is its overall value. For commercial income producing buildings the value of the building is generally tied to the income stream, mainly the net operating income (NOI).The NOI is potential gross income, less building operating expenses, vacancies, and collection losses. The value of the building can then be estimated or calculated by using a revenue multiplier or capitalization rate of the building's NOI. If the building has been operating for a while and has a history of NOI, another approach called discounted cash flow can be used. This technique uses discount rates to projected cash flows or income and calculates a present value of the building. So income, expenses, occupancy, and rental rates all play a part of the valuation of the asset and determine the major metric of the financial performance of the building.

2.2   Security and Life Safety

Buildings must provide for the physical protection of its occupants and assets. This includes protection from crime, vandalism, espionage, fire, accidents, and environmental elements. Typically a security threat assessment of the building is conducted and then measures are identified and deployed. These may include deterring, detecting, delaying, mitigating or notifying of any attempt to injure, damage, modify, or remove an asset or person. These measures are put into place in three ways: the building systems, architectural countermeasures, and security operations. The building systems needed to provide security and life safety include fire alarm, video surveillance, access control and intrusion detection. Architectural countermeasures include landscaping, doors, lighting, vehicular standoffs, and air intake (protecting building occupants from airborne chemical, biological, or radiological attacks). Building security operations include emergency preparedness, training, policies, patrols, and so forth. So the security performance of a building is based on regular threat assessments of the building, optimal operation and testing of the security related systems, prudent use of architectural countermeasures and a fully functional security operation. Security and life safety are affected by many different factors such as location and age of the building, composition of the building occupants, climate, economic conditions, and education levels. Data on the number and type of security incidents in the building, trends in incidents, crime statistics for the area, and occupant surveys can be used to evaluate the security performance of a building.

2.2.1   Operations and Maintenance

Efficient and effective operation and maintenance of a building is critical to its performance. It includes a variety of tasks: space planning, work order processing, energy management, asset management, management responsiveness, systems performance, provision of specific space needs and

Figure 2.1   Commercial building.

requirements, cleaning, training of staff, management of contractors, and so forth. Data from organizations such as BOMA and the International Facility Management Association (IFMA) allow building owners to compare the operations of their building with other comparable buildings to evaluate building performance. For example, IFMA can provide information on gross and rentable square footage by industry type and facility use, square footage per occupant, janitorial costs and staffing, maintenance costs, roads and grounds costs, utility costs, life and safety costs, emergency and disaster planning costs, facility management information technology costs, employee amenities costs and best practices. So one way to evaluate the operational performance of a building is to gather operational data on the building and compare it to the operational metrics of similar buildings.

2.3 Productivity and Satisfaction of Building Occupants

Buildings enable their occupants to work, play, meet, shop, sleep, eat, socialize, educate, learn and a host of other activities. One performance criteria of a building should be how well it succeeds in enabling its occupants. This involves the comfort of the occupants, both physically and psychologically. The physical part is straightforward, involving thermal comfort, appropriate lighting and air distribution, workspace layouts and the technology systems to to make the occupants' tasks easier. These technology systems might include systems for digital signage, Wi-Fi, in-building cell phone coverage, asset location systems, audio video systems, and so forth. The physiological effect may relate to the building's image, appearance, and aesthetics.

Another measure of building performance is the productivity of building occupants. The largest effect on productivity over the last several decades has been the penetration of IT technology and the Internet which reduces the time to access information, perform tasks and communicate. The workspace itself also plays a very important role. In a workplace survey conducted by a major architectural company, they found that effective workplace design directly correlated to improved business performance. The architect created a Workplace Performance Index that accounted for the criticality of the work mode, the time spent on the work and the effectiveness of the space for the particular work mode (work modes are activities such as collaboration, focus and concentration, learning and socializing). This study accounted for the physical attributes of the space: layout, lighting, air, storage, furniture and access/privacy. What they found was that top-performing companies had workplaces with higher performance indices. They also found employees in workplaces with higher performance indices had greater job satisfaction, organizational commitment and were more engaged with their employer, all very positive for the businesses and organizations. The metrics and methodology of evaluating the satisfaction and productivity of building occupants have been developed and at the core it is a survey of people that use the building. The feedback from those people, whether they are office workers, shoppers or teachers is valuable input to enhance building operations or improve the design of the next building.

Lighting in workplaces can affect our disposition, satisfaction and well-being and is important to occupants comfort and productivity. Natural light is the best light for work but many commercial offices don't have enough daylight. Lighting systems are used to complement or supplement daylighting. The quality and quantity of such artificial lighting is key to comfort and productivity. Occupants also want control of the lighting in their workspace or desk, enough lighting to carry out their tasks, and the light to provide a pleasing ambience.

Many times new commercial buildings install generic lighting in tenant space prior to knowing who the tenant may be and the eventual tenant is left with general lighting rather than lighting specific to their needs. Issues such as lighting control, quality of the light, lighting too high or lighting at the wrong locations occur. Also, over time building space may change or be renovated and it's at that point that lighting needs to be reconsidered.

In the past, offices in a commercial office building had four walls and a door. Over time many offices became desks in open plan spaces with the system furniture, where desks and cubicles were interlocked. The intent was primarily to facilitate employee interaction and collaboration. In some cases the objective was to reduce the space needed as well as cost. Many current office environments may have a few private offices and use flexible assignment of desks or offices via *hoteling* or *hot-desking*. Large open plan spaces can create acoustic issues; loss of privacy , and high noise levels, all of which cause distractions and decreased productivity. Studies show office workers biggest complaint is privacy, specifically overheard conversations.

To address acoustic issues in office spaces owners or tenants have implemented several initiatives. One strategy is simply to separate incompatible office uses. That is, rooms meant for interaction, collaboration and a level of privacy, such as conference rooms, huddle rooms, focus rooms, etc. would be located separate from open office space. The same would be true of support rooms, such as coffee bars or copier rooms.

If partitions are used in open office spaces, they should have a low noise reduction coefficient so the partition does not reflect speech when an occupant is seated at their desk. Another option is the use of sound absorbing ceilings and walls. Sound rated walls should be used for training rooms, conference rooms, and executive offices.

Another option to address acoustic issues in office space is the use of sound masking systems. These systems matured in the 1970s with the use of pick noise which could match the frequencies of human speech and improved audibility. A sound masking system uses a series of loudspeakers installed in a predetermined grid pattern in the ceiling. There are methods to control and configure grid zones and output to optimize different spaces. The system basically distributes an engineered background sound throughout a work space. Independent studies have indicated that sound masking systems improve worker productivity, reduce stress, and increase job satisfaction

One of the most important factors affecting occupancy satisfaction is thermal comfort. If you were to ask a facility engineer what the most common service call is from a building occupant he would tell you it's "hot and cold calls". These calls are driven not only by the air temperature but also air velocity, relative humidity, and the temperature of the occupant's immediate surroundings. There are a number of other factors that can affect an occupant's thermal comfort: their clothing, physical activity level, overall wellbeing, and food and drink.

While we focus on thermal comfort and health, thermal aspects of the building must also prevent mold and mildew as well as damage to the building's materials. Thermal comfort in a commercial building must adhere to ASHRAE Standard 55 Thermal Environmental Conditions for Human Occupancy. This essentially defines thermal comfort in commercial buildings. This standard provides methods to determine thermal environmental conditions (temperature, humidity, air speed, radiant effects) for buildings in which a significant proportion of the occupants (80%) will find acceptable comfort at a certain metabolic rate and clothing level.

Short of a quantitative whole building performance methodology, building owners will need to use comparable data, interviews, observations, surveys, tests, and demographic and financial data to evaluate other building performance factors. Note that there may be other measures to consider in a given building's performance than the ones covered here and other methodologies to develop those performance metrics. One option of measuring performance is the use of Post Occupancy Evaluations (POE). These evaluations are primarily geared towards the occupants, but also include building staff, visitors, and others regarding how the building meets user's needs and suggestions for improving building performance. A POE usually involves both quantitative and qualitative aspects. The evaluation can be done as a focus group, via an interview or a questionnaire. Other information, such as energy consumption or lighting levels, air quality, or the number and responses to work orders can be taken into account. While POE obviously involves post occupancy, it can be used prior to the project to identify project success or set a baseline.

It is difficult, if not next to impossible, to quantify all the different aspects of a building environment on occupant productivity and satisfaction. However, a number of studies on a conducive building environment indicate modest, measurable improvements in productivity which certainly result in significant financial gains.

Essential Attributes of a Smart Building

3.1 Cabling Infrastructure, Lighting Control Systems, and Facility Management Systems

Building owners, engineers, designers, contractors and facility management organizations are all aiming to build new buildings or renovate existing buildings identified as smart buildings. In general we think of smart buildings as being innovative, using advanced technology and materials, contributing to reduced energy usage, improving the building's sustainability, and providing more efficient and effective operation. But the high-level understanding of smart buildings doesn't do justice to the concept; a catch phrase doesn't assist design teams or contractors, and it really doesn't reflect the complexity of today's buildings.

Generally, ethereal attributes describing smart buildings don't help in defining a smart building so that the industry can have a common understanding of the concept. In contrast, look at green buildings or energy efficient or sustainable buildings; here you'll find dozens of countries with environmental or energy building certifications with specific details on the requirements to be certified as a green building. Recall twenty years ago when the United States Green Building Council (USGBC) was formed, the idea was to provide the building industry with a system to define and measure green buildings. We need a similar effort for smart buildings to support the building industry.

There have been various attempts to define a smart or intelligent building. They include the Asian Institute of Intelligent Buildings (AIIB), the Building Research Establishment Ltd., the Intelligent Building Society of Korea (IBSK), the Shanghai Construction Council (SCC), the Architecture and Building Research Institute, the Ministry of the Interior of Taiwan and finally the Smart Buildings Institute (SBI) in the USA. (http://www.smart-buildingsinstitute.org/)

The SBI has certified buildings in the Americas as well as internationally. The SBI certification structure is similar to LEED, having prerequisites, with other credit-based measures where applicants earn points during construction and operation of the building. SBI also developed educational content that really isn't available elsewhere, such as advanced automation, system integration, and so forth

Defining the attributes of a smart building cannot be done in few words; buildings are too complex and the features of a smart building too numerous. The following sections will frame the major attributes of a smart building and provide details.

3.1.1 Cabling Infrastructure

Cabling, cable pathways, and equipment rooms are important because installed cable tends to be in the building for 20-25 years. What is initially designed and installed is vitally important. Cable may seem mundane, but, with constant advancements in cable throughput, reductions in potential interference, and increased low voltage power over the telecom cable it means the long lifecycle requires the installation of the latest in cable technology.

The smart buildings approach is to maximize the use of structured cable plant for multiple building systems. A smart building will also supplement the wired infrastructure with the use of wireless systems. As IT continues to penetrate a building this infrastructure becomes increasing critical and a key attribute of a smart building.

3.1.2 Lighting Control Systems

Lighting affects several important aspects of a building. There is the aesthetic impact it will have in the building and its spaces, resulting in how occupants will perceive and react to the built environment. There is the more practical and functional aspect of providing lighting to facilitate the work occupants undertake in and around a building. And finally, lighting consumes significant energy.

A smart building entails a centralized lighting control system in all the building's usable square footage. The system should provide a dimmable lighting strategy. It also should have a global and a zone light schedule and override capabilities.

3.1.3 Facility Management Tools

The operation of a facility represents the longest duration of the lifecycle of a building and the largest portion of the total lifecycle cost. The facility management responsibilities are wide-ranging and multidisciplinary; varying from responses to immediate alarms to regular maintenance work orders or long term capital planning. Given the fact that today's buildings are increasing more complex, the need for a smart building to have facility management top of the line tools and processes to efficiently manage the facility becomes essential. The software facility management applications shape the administrative processes within facility management. They include: a work order system, a preventative and predictive maintenance application, space planning management, inventory control system, and asset management.

BIM Integration

Building Information Modeling is a significant data management tool for new construction. The data and modeling generated in the use of BIM should be exported in the facility management systems.

3.2 System Integration, Audio-Visual Systems, and Water

3.2.1 System Integration

Integration functionally links two building systems to provide functionality that neither system could by itself. The best example of this process is the integration that takes place with the fire alarm system. The fire alarm triggers the HVAC system to control smoke and ventilation, the access control system to provide egress for occupant evacuation, and the elevator cabs are automatically recalled to a facility's predetermined floor of egress. Without the automated systems' integration each of these components would have to be separately and manually adjusted. The theory is that the whole is greater than the sum of the parts. Another reason to integrate systems is to combine the system data. Rather than simply looking at data from one building system a database with data from multiple systems is created so that holistic data can be analyzed, correlated, and insightful building metrics can be developed.

3.2.2   Audio-Visual Systems

Many times the deployment of audio visual systems are undervalued and marginalized. However, AV systems can affect tenant satisfaction, energy usage and operations, and are an important part of a building. The AV industry is interesting because contractors, manufacturers, AV managers, and technicians have hands-on experience with system integration, albeit at a different point than the integration issues building owners and facility managers generally deal with. Also they have experience with the design and construction process. Typical AV integration may include IT, shading, security, and other building systems. There are several AV facets and options that are needed to make a building smarter:

3.2.3   Paging and Messaging Systems

A building's paging and messaging systems may be thought of as secondary systems, but often they are a part of life safety. Paging is also used in many educational, large healthcare, corporate and government buildings such as airports, and convention centers. Legacy paging systems were typically a separate piece of equipment connected to a PBX (something which is now long gone). Current systems are built on VOIP-type technology and IP end devices, including IP speakers. The network connectivity allows authorized users to broadcast audio simultaneously to speakers and IP telephones. Some systems have the capability to concurrently send a multicast audio stream and text messages that can be delivered to not only paging speakers and IP telephones but also PCs, tablets, and smartphones.

Some messages could include information on weather conditions or event reminders for organizations. More importantly, the messaging and notification can be used for emergency situations. The notification capability of the system can provide real-time communications within a building or across a campus regarding an emergency situation thus significantly improving life safety and security.

Digital Signage

Digital signage is a very compelling technology. The medium has stickiness—few people will not at the very least, glance at or pay attention to a plasma or LCD display. It's a communication system that's effective, immediate, and dynamic. It can be utilized in a variety of building types. Digital signage can be used to inform, entertain, communicate, advertise, and enhance the experience one has within a building. Different building types will have various uses for digital signage.

3.2.4 Water

Water is a critically important resource; maybe our most precious resource. Where alternatives exist for energy sources there are no alternatives to water. If you don't think water is important consider that people can survive longer without food and certainly without energy than they can without water; not surprising considering that two thirds of the human body by weight is composed of water. Water also has a direct connection to energy use. The critical nature of water is recognized in green building programs throughout the world.

3.3   Occupant Satisfaction, Fire Alarm, Networks and Security

3.3.1   Occupant Satisfaction

Buildings enable their occupants to work, play, meet, shop, sleep, eat, socialize, educate, learn and a host of other activities. This involves their comfort, both physically and psychologically. Considerations such as thermal comfort, appropriate lighting for the occupants' activity, control of lighting and air distribution, workspace layouts and the technology systems make it easier for the occupants to complete their tasks. The physiological effect may relate to the building's image, appearance, and aesthetics.

3.3.2   Fire Alarm

The primary job of the fire alarm system is to warn building occupants of a fire so that they can safely evacuate the premises. The fire alarm system is one main component of a larger fire protection system. True fire protection encompasses mechanical systems, electrical systems, structural attributes, and architectural aspects such as means of egress. These may include: fire sprinklers that discharge water when a fire has been detected or a predetermined temperature has been exceeded. The structural and architectural aspects of the building play a part in fire protection. The building must be fire-rated and have a fire resistant structure, as well as intuitive means of egress for occupants.

3.3.3   Network and Security

The building automation industry now has legitimate and reasonable concerns regarding the security of building control systems, especially in smart buildings where advanced technology is deployed. This apprehension is amplified in newer buildings because of increased penetration of IT infrastructure in building control systems. The overarching security concern is primarily about network security and less about physical security, although the two are certainly related.

3.4   Electrical, Building Metering, and Video Surveillance Systems

3.4.1   Electrical

Without stable, high quality electricity our world would be radically and negatively impacted. We rely on safe and reliable electricity from our utilities and within our buildings.

A smart building has an electrical infrastructure that is dependable, adaptable, and cost-effective to maintain and operate. In addition to the electrical infrastructure building owners also must have a relationship with their supplier for two-way communication with the utility grid, demand response notifications, and optimal rate structures. Building owners may address the use of renewable sources or a microgrid for a building or development that can locally generate electricity. A prerequisite for electrical distribution is a power monitoring control system, able to examine the quantity and quality of power being supplied and consumed. Measurement produces data, and data is the underpinning or basis of control and management. Electrical backup capabilities are also a required for smart buildings, especially critical facilities such as healthcare and data centers.

3.4.2   Building Metering

Smart meters gather data in real-time and provide not only usage data but also information on outages, malfunctions, and quality. There are generally two types of power meters for buildings. One is for monitoring and measuring power to the whole building. The second type is a submeter, meant to measure usage for specific spaces, tenants or electrical circuits.

There is also interest water monitoring and management systems consisting of water meters, sensor-operator water fixtures such as faucets, urinals, water closets, occupancy sensors, automated ball valves and water valves.

3.4.3   Video Surveillance Systems

It's a given that any sizable building will have a video surveillance system. Video surveillance systems, also known as closed-circuit television systems (CCTV), are one part of a facility's larger security and life safety plan. The larger plan may include physical and operational aspects of security as well as other security or life safety systems, such as access control and intrusion detection. Much like the broader electronics marketplace, the technology for video surveillance has changed from legacy analog to digital technology.

IP-Based video surveillance leverages existing IT infrastructure and contributes to lower cost of installation.

Cameras can detect smoke or fire, identify specific individuals, detect motion, determine if objects have been moved and provide occupancy data including the actual number of people in a given space. Generally, if you can develop a pixel template of the event or condition you are trying to track, the video analytic software can detect the event or condition.

The array of consistent analytic tools related to video cameras are extensive. They include:

▶ Facial Recognition

▶ Motion Detection

▶ Missing Objects

▶ Reading License Plates

3.5   Advanced Building Management Systems, Communication, Data Infrastructure and HVAC, Access Control and Sustainability

3.5.1   Advanced Building Management Systems

The driver in advanced building management systems is the increasing complexity of buildings. From an equipment or hardware perspective we now have buildings with energy and sustainability systems which are relatively new for buildings, systems that even five years ago were not commonplace. These include systems such as rain water harvesting, exterior shading, water reclamation, electric switchable glass, sun tracking systems, and so forth. Maintaining and optimizing each of these new systems is a challenge, further burdening and challenging facility management. The other aspect of increased complexity is related to management decisions regarding building operations. Many of these decisions now involve several variables, with some situations requiring real-time decisions, for example a demand response event from a utility requiring immediate action.

The shortcomings of the typical legacy BMS is quite a long list, including limited integration capabilities, inadequate and elementary analytic tools, proprietary programming languages, a dearth of software applications, and legacy user interfaces.

A smart building will have an advanced building management system with an open programming language where all integration is accomplished via software. It requires middleware to normalize and standardize all data from subsystems into an open, standardized database using SOAP/XML or other computer software exchange architecture. The database would include all physical, virtual and calculated points. The user interface to the advanced systems include displays and dashboards completely configurable and customizable by users, with access via a browser. The system would be capable of data exchange with information in enterprise and business level software, providing a suite of software applications such as energy management, building performance analytics, alarm management, and automatic fault detection and diagnosis.

3.5.2 Communication and Data Infrastructure

The method of communication and data exchange within and between building systems is vital. It is a foundation that will determine the difficulty or ease of integrating system functionality and data. Smart buildings shun proprietary protocols in favor of standard open communications protocols based on the ASHRAE BACnet I/P, Lonworks, OPC DA, Modbus TCP, oBIX, XML, SOAP and SNMP standards of data exchange or similar open standard protocols. Many building products now incorporate open protocols, some going through a process that verifies or certifies their adherence to the protocol standard.

The network architecture of building systems should take into account the minimum speeds for serial buses, maximum size in points and devices per serial bus, maximum number of serial buses per network controller, and the use of native open-protocol controllers versus gateways in existing buildings.

The adherence to standard open protocols and detailed network design extends not only to building control systems but also to facility management systems, business systems, and IT systems in the use of standard database structures such as SQL or ODBC, Oracle or DB2.

3.5.3 HVAC

In many ways, HVAC equipment is the most complex building system, with numerous components arranged to produce heating, cooling and ventilation through the principals of thermodynamics, fluid mechanics and heat transfer.

The HVAC system not only makes the building comfortable and healthy for its occupants, it manages a substantial portion of the energy consumed, as well as plays a critical role in life safety. In maintaining the building's air quality the HVAC system must respond to a variety of conditions inside and outside the building (including weather, time of day, different types of spaces within a building and building occupancy) and do so while optimizing its operation and related energy usage. Given the variety of conditions and the potential complexity of a substantial HVAC system this necessitates extensive automation and system integration. For example, in a smart building we expect the HVAC system to automatically sequence chillers, pumps, and boilers, as well as automatically rotate parallel chillers, pumps, and boilers by accumulated run-time. The HVAC system should also perform an optimal start calculation based upon real occupancy history instead of estimated start times.

Control of the HVAC system for occasional use facilities such as meeting rooms, conference rooms, and cafeterias is also important in reducing unnecessary energy consumption. HVAC for those spaces needs to be integrated into another system which can supply data to the HVAC system regarding use or occupancy. These include data from an access control system, video surveillance, a people counter system, lighting control, a RTLS/RFID system or more likely, occupancy sensors.

The HVAC system also plays a substantial role in a demand response events as well as demand limiting. Data from the utility or a power management system communicated to a BMS and chiller controls can be used to adjust the electric demand of the HVAC system to an acceptable level.

3.5.4 Access Control System

Access control systems are a critical component in smart buildings as security has become more important. The access control system is also essential for life safety and is interfaced to the fire alarm system to facilitate building egress during life safety evacuations. Access control systems must interface or integrate with several other smart building systems (video surveillance, HVAC, and others) as well as share data with business systems, such as human resources, time, and attendance.

In a smart building, one electronic access control system for non-public areas should be deployed. Within secured areas the access control system would provide two levels of authentication. The system should support of-fline operation to allow doors to function if network connectivity is lost. The access control system should be supplemented by an intrusion detection system at potential unauthorized entrances, such as windows. While access cards are generally used in many systems, biometric authentication may be utilized for an additional degree of security.

The access control system should be configured to maximize security. For example, its use of vertical transport systems (elevators) can provide selective access to floors based on occupant identity, as well as spaces such as parking and garages.

Security levels would be determined by individual, floor, or areas, and access privileges can be changed in response to building occupancy states (i.e. time of day). An access control system can also generate anonymous occupancy statistics for building spaces and zones. Such data can be used to correlate occupancy to other building systems such as energy consumption or lighting schedule. One of the largest problems with access control systems can be *piggybacking and tailgating*. Piggybacking happens when someone with legitimate access to a building allows someone without access to the building to come in with them. Tailgating involves taking advantage of someone who legitimately has access to the building, where a trespasser enters the building with a person (or group of people) without their knowledge. One way to prevent piggybacking and tailgating is a *mantrap shield* that use sensors to ensure that only one person is entering the building using one set of credentials. Mantrap shields can also be configured with separate compartments so that if more than one person is sensed passing through the first door, the second door will remain locked.

3.5.5 Sustainability and Innovation

Green buildings and smart buildings have different focuses but they also overlap. A primary component of a smart building is energy efficiency and sustainability, acknowledged by an industry certification such as LEED, and a clear policy and plan for energy management by the building owner. This plan may involve delegation of responsibility for energy consumption, as well as tracking, monitoring, and reporting systems for energy consumption. In addition, building owners would participate in demand response and automated load shedding in cooperation with the utility company

Innovation is integral to a smart building. Innovation by the building owner, designer, contractor or manufacturer that can demonstrate benefits, value and exceptional performance should be recognized and incorporated into the methods or criteria for deploying a smart building.

3.6 The Constantly Evolving Smart Building

Technological progress is like an axe in the hands of a pathological criminal
—Albert Einstein

The smartest parts of a building, its systems and materials, are driven by innovation and technology. They may have focus on long term building operations and performance, but, some are simply major game changers. Inventions such as elevators, construction cranes, and power tools are examples of equipment that changed the way buildings are designed and constructed. While electricity is not a human invention, the commercialization of electricity and its use in buildings was a milestone. One of the latest game changers has to be Building Information Modeling (BIM); designing, fabricating, implementing and managing construction in multiple dimensions.

The backdrop for buildings and related automation is now tied to the innovative information and communications technology for building systems, building design, construction, operations and building occupants. Many astute building owners embrace technology and innovation. However, it can be a challenge and uncomfortable for some building owners, architects, engineers, and facility managers to innovate and change. It's always easier to keep doing what you've been doing than to put forth the effort to do something different. Some may choose not to innovate based on the rationale that they can mitigate risk. Nevertheless the smart designers and engineers will examine innovations, assess risk, and gauge the ultimate value for building owners, as well as include new ideas, systems and products in discussions of a project concept.

The broader environment for smart buildings is related to: (a) the habituation of the global society to communications and information technology, and (b) the emergence and convergence of three related markets: smart buildings, smart cities, and the Internet of Things.

The morphing of these different markets has happened very quickly and almost without any measured intent of the building, city, and IoT industries. What's driving the transformation is the commonality among the three markets. Each of these entities are pushing the envelope to advance the experience and performance for building occupants, citizens and individuals, using technology as the enabler. The tools being utilized to provide that experience are similar. They include system integration, the acquisition and management of data, the analysis of data, the creation of new software applications, the development of performance metrics, and the visualization of the data tailored to the entity or individual consuming the data.

3.6.1 Smart Buildings and Cities

The best and most obvious example of these markets evolving is smart buildings and smart cities. The catalyst for smart cities is population growth, with the population becoming more urbanized. Predictions are that seventy percent of the world's population will be in cities by 2050. Urbanization spawns buildings and requires building owners and the city community to take responsibilities for sustainability, energy management, and livability. Here's a short list of mutual issues for cities and smart buildings:

▶ Energy: Cities need secured and adequate energy supplies. Energy also affects the environment of a city and the cost of living. The city's utility grid needs to communicate with the cities' buildings, and the power grid. Individual buildings should acquire, analyze, and share energy data. Major buildings and large developments should be encouraged to develop microgrids with a variety of energy sources to improve reliability, stability, and efficiency. Communication between the grid and the buildings allows for demand response, but, also sets up the microgrids as energy sources especially when the grid's capacity is being taxed.

▶ Water Distribution Systems: Humans can live without energy; but they can't without water. Water is our most precious resource so cities as well as building owners need to undertake water conservation and waste water treatment. Much like energy monitoring, real-time monitoring and management of water should be required, not only for consumption metrics, but also for leak detection.

▶ Transportation: Traffic congestion and a lack of alternative transportation modes are major negatives when it comes to a city's livability and its economy. Traffic oftentimes is the top complaint citizens have about their city. Cities need to deploy intelligent traffic systems such as traffic signal control systems, license plate recognition, and real-time data from other systems to utilize predictive analytics in reducing travel times. However, it is critical that the road system be supplemented by alternative modes of transportation. Large developments and buildings are typically part of the transportation plan, with some cities requiring building owners to have a transportation management program and a plan for trip reductions in order to reduce traffic and parking loads.

▶ Public Safety: Safety and security is key to a city and its buildings. For the city it involves multiple agencies and organizations; police, fire, emergency, courts, neighborhood groups, and more. The city is in a position to collect intelligence, use predictive data analytics, communications and situational awareness to help in predicting crime areas.

Building owners are generally proactive and do a security risk assessment which involves systems to protect building occupants, resources, the building structure, and continuity of operations. While a city and a building owner have the same security and safety goals, their scope is different but complimentary.

▶ Digital Services: Cities as well as buildings need to deploy electronic and internet services. This may be smartphone apps for finding your way in a building or a city, as well as interacting with a city official or a building manager electronically, essentially providing convenience.

3.6.2   The Internet of Things and Smart Buildings

The Internet of Things (or everything) is somewhat of a wild card. It seems to be a concept with little definition. The Internet of Things will basically connect everything to everything else using the internet, which will lead to an unprecedented level of automation for a variety of fields. The fact that IoT a concept without boundaries may be a good thing and spawn new ideas. However, standards development may be the Achilles heel of the IoT. It may take some time and may eventually result in multiple standards, including proprietary ones from groups of commercial technology companies. This could delay the IoT market everyone is expecting.

This level of connectivity could provide integration to enhance functionality that none of the systems or devices could provide individually. Or it could simply acquire data from devices and analyze or mine the data in order to develop and gather information. This is essentially what many astute facility management groups are already doing; integrating building systems to provide greater functionality and deploying analytic software applications to improve the performance of building systems. It's difficult to see if IoT can add much for building owners who already are integrating and analyzing data.

If you examine the commercial companies involved in the various associations or consortiums attempting to create the IoT standard, all are technology companies; chip manufacturers, and hardware, telecom, and software vendors. It would seem that home automation, wearable technology and IT will be significant sectors of the IoT given the involvement of IT companies.

What effect or influence will the IoT really have on building management? If the initial products are related to wearable technology and home automation there will not be much interest by facility managers unless a building is mixed use, in which some apartments or condominiums require home automation to be deployed. Facility management can and will deploy additional sensors if needed without thinking much about IoT. With the building automation industry's long history of a handful of well-known global communication protocols and the excellent gateways and middle-ware in the market, facility managers have the tools to take their buildings to a higher and more valuable level of building automation, with or without the IoT.

Numerous research companies and organizations are predicting significant growth in the smart building market. One company, Markets and Markets, expects the global marketplace to grow at a stunning compounded annual rate of 35% over the next five years. While it's difficult to compare market research because there are various definitions of what submarkets (smart homes, smart grid, etc.) make up the larger smart building market, most research in the smart building area indicates substantial growth world-wide. Given that the general economic outlook for the global economy is relatively flat, or for modest growth at best, it's impressive that smart buildings are such a rapidly growing segment, but, this is by no means a surprise.

There are several factors creating demand for smart buildings. One of the most potent is the results from building owners that have already deployed smart building technology. These building owners have found reductions in energy consumption, enhancements to operations, and a very attractive return on investment. Such examples and stories validate the approach, verify the likely results, and reduce the risk for other building owners to plan to deploy the technologies. Another element driving the market for smart buildings is our global society's habituation to real-time information and communications technology; people not only accept cutting-edge technology as an integral part of our buildings but expect that their buildings will be smart.

An additional factor to consider is that while the marketplace is rapidly evolving, there are existing elements that comprise a smart building. These include system integration, advanced building management tools, extensive automation and sensors, energy management, enterprise data management, data analytics, software applications and the leveraging and incorporation of IT. It is this emerging clarity that can guide designers, contractors and manufacturers, as we complete the definition of a smart building by addressing HVAC, communication and data infrastructure, access control systems, advanced building management systems and sustainability.

Information Technology in Building Systems

4.1 Overview

Transformative periods in the building industry occurred several times in the 20th century with the introduction of mechanisms such as plumbing, construction cranes, and elevators. Thirty years ago, just prior to the mass introduction of personal computers for businesses, the level of technology in a building was meager. It consisted of the local regulated public telecommunications utility installing services in a building, a mechanical contractor installing a pneumatic control system for the heating, cooling, and ventilation system, a fire alarm system, and maybe a dedicated word processing system. While we've come a long way since those days, we're still in a very early stage of fully deploying and integrating sophisticated building technology systems.

In due course, buildings will become full of technology. Walls and ceilings will be embedded with sensors; every aspect of a building's performance will be metered and measured; software tools will be used to automatically optimize building systems without human intervention;

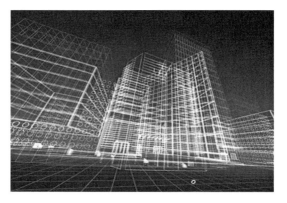

Figure 4.1   The use of information technology to design a building in 3D.

real-time information on the building will be provided to occupants and building management relevant to their particular needs; buildings will be fully interactive with the power grid; cars will be efficiently parked via conveyers, and geo-spatial location systems will be deployed for every building asset. Facilitating this transformation is information technology.

4.2   Communications Protocols

There is an increasing role for information technology in our lives, including building control systems and facility management. All major communications protocols used in building systems (BACnet, Modbus, Lonworks, etc.) now have a version for Internet Protocol (IP) or Transmission Control Protocol (TCP) that allow the BAS protocol to ride on an IT network. We now have building management systems; basically an IT device with a server, IT operating system, IP address, and IT database. Further evidence of the IT infiltration includes international standards for cabling related to building automation systems that are identical to that of IT, BAS controllers using Wi-Fi, and the current focus on data analytics for building system data. This IT infiltration has and will cause disruptions and adjustments between IT and Facility Management organizational roles.

4.2.1   Wireless Infrastructure

The use of wireless networks control systems in buildings has many advantages. By eliminating the need for cable and related conduit the initial costs for deploying system sensors, meters, and control devices are reduced and installation time is shortened. Wireless is the ideal approach for retrofitting existing buildings where the lack of cable pathways is an impediment.

The major difference between the performance of wired and wireless networks is network communication capacity or bandwidth. However, building systems field devices generally do not need much bandwidth and conduct their business at relatively low data rates. Another potential issue with wireless is that some wireless transceivers may use batteries which require regular replacement. The low data rates and the possibility of periodic battery replacements are minor strikes against wireless compared to its flexibility, cost advantages, and reduced installation time. What follows are some of the choices for wireless networks for building control systems as well as some of the building devices that can be deployed wirelessly.

4.2.2 Wireless Network Types

There are wireless networks using licensed radio frequencies and unlicensed radio frequencies. There are wireless technologies such as Zigbee, EnOcean, Z-Wave, Wi-Fi, RFID, Insteon, Bluetooth, etc. The following is a snapshot of some of the major technologies:

4.2.2.1 Zigbee

Zigbee is a wireless technology standard (IEEE 802.15.4) which provides for low data rate networks. It uses unlicensed frequencies (900 MHz in the US, 868 MHz in Europe, and 2.4 GHz worldwide) which are also available for cordless phones Wi-Fi, and other devices. The standard is aimed to address residential, building, and industrial control devices. It is specifically useful for sensors and control devices of building automation systems within a smart building where very small amounts of information or data are being transmitted. Zigbee also has uses in home automation, industrial automation, home entertainment systems, and smart meters.

The maximum speed of Zigbee devices varies up to 192-250 Kbps (a measure of bandwidth, kilobits per second). The maximum distance varies between 20 and 50 meters. Zigbee has several advantages: (a) low power usage since the devices only require two AAA batteries, (b) wide support from more than 100 companies that support the standard (Motorola, Honeywell, Samsung, Mitsubishi, and others), (c) mesh technology which allows Zigbee, like Wi-Fi, to be configured in several topologies including a mesh topology allowing multiple transmission paths between the device and the recipient, and (d) system scalability where thousands of Zigbee devices can deployed within a building.

4.2.2.2 EnOcean

EnOcean Alliance is a consortium of companies in North America and Europe that are developing and promoting self-powered wireless devices. The initial consortium formed in 2008, and includes Texas Instruments, Omnio, Sylvania, Masco, and MK Electric. The consortium claims the largest installed base of wireless building automation networks.

The main objective of this technology is to allow sensors and radio switches to operate without batteries reducing maintenance. To do this the technology uses *energy harvesting* which exploits slight changes in motion, pressure, light, temperature, or vibration to transform small energy fluctuations into usable electrical energy. The devices transmit at 120 Kbits per second, up to 300 meters with a data packet of 14 bytes. The transmission frequency used for these devices is 868.3 MHz, an unlicensed radio frequency.

EnOcean GmbH, a spin-off company of Siemens, supplies the transmitters, receivers, transceivers and energy converters to companies such as Distech Controls, Zumtobel, Omnio, Osram, Wieland Electric, Peha, Thermokon, Wago, Herga and MK Electric, who then develop and manufacture products. While the products tend to focus on building automation they are also targeted for industrial automation and automotive markets.

EnOcean is based on an international standard, the International Electrotechnical Commission (IEC) standard—ISO/IEC 14543-3-10—for wireless applications with ultra-low power consumption. They have created EnOcean Equipment Profiles (EEP) for EnOcean devices that ensure interoperability of different end-products based on EnOcean technology; resulting in equipment from one manufacturer being able to communicate with equipment of another manufacturer.

4.2.2.3 Z-Wave

Z-Wave is about wireless solutions for residential and light commercial applications. The Z-Wave Alliance is an open consortium of over 160 manufacturers. Members include Cooper Wiring Devices, Danfoss, Fakro, Ingersoll-Rand, Intermatic, Leviton, Universal Electronics, Wayne-Dalton, Z-Wave and Zensys.

At the core is the Z-Wave protocol, developed by Zensys, a division of Sigma Designs. Sigma Designs provides embedded networking software and Z-Wave chip solutions for manufacturers and OEMs. The Z-Wave protocol stack is embedded in the chips, and flash memory is available application software. The standard is not open and is available only to Zensys/Sigma Design customers.

Z-Wave operates as a mesh network in the 900 MHz radio frequency range and is optimized for low-overhead commands such as on-off (as in a light switch or an appliance), with the ability to include device metadata in the communications.

Each device on the Z-wave network has an individual code or address. A single Z-wave network supports up to 232 devices. Multiple Z-wave networks can be combined via gateways. The controllers can be handheld remotes, wall panels, or an internet interface via a browser. Like some of the other wireless networks, Z-Wave is a mesh networking technology where each node or device on the network is capable of sending and receiving control commands. Z-Wave can also use power line communication technologies.

Z-Wave has a speed of 40 Kbit/s with a range of about 100 feet or 30 meters. The radio frequencies used include 900 MHz ISM band: 908.42MHz (United States), 868.42MHz (Europe), 919.82MHz (Hong Kong) and 921.42MHz (Australia/New Zealand).

4.2.2.4   RFID

Radio-frequency identification (RFID) is different than the other building wireless systems. It can't control anything, and it only identifies things. Its primary use is in asset management and security. RFID tags are incorporated into products or carried by people to identify and track their location using radio frequencies. RFID is deployed in retail, hospitals, airports, education, and other building uses.

Systems generally consist of RFID readers and tags. RFID tags are simply radio transponders. They are a small integrated circuit or computer chip which has a tiny radio antenna built in. In passive RFID systems the tag does not have its own power source, the tag absorbs energy from the system reader antenna, a process called *coupling*. The tag is programmed with a unique identification. When the tag is excited by and absorbs the radio waves of the reader antenna it sends out its unique ID which is picked up by the reader antenna.

Active RFID systems tags have their own power source and don't need to use the reader's antenna radio waves to power up and transmit their identity. Active tags have greater range, can store larger amounts of data, and are larger than passive tags.

RFID tags come in various sizes and shapes to address a variety of uses. The tag can be paper thin to fit inside a book. They also can be directly mounted onto equipment, embedded in wrist straps, attached to clothing, or worn on a belt.

Wireless tracking systems are only as good as their networks. RFID readers have an antenna attached to them. Essentially the reader interfaces or sits in between the wireless portion of the system (the antenna) and the head end or host system. The antenna attached to the reader sends radio signals out to activate tags. It listens for tags to communicate and once a tag responds, reads the data transmitted by the tag and sends it to the reader. Readers can have multiple antennas attached. The reader can decipher the signal and send the data to the host server.

RFID operates in several radio frequencies: 125 kHz or 134 kHz low-frequency systems, 13.56 MHz for high-frequency system, and 2 or 3 frequencies for ultra-high frequency systems.

4.2.2.5 Wi-Fi

Wi-Fi basically replaces a cabled Ethernet connection with a wireless device. Current Wi-Fi systems operate in two unlicensed radio frequencies, 2.4 GHz and 5 GHz. The Institute for Electrical and Electronics Engineers (IEEE) has set four standards for Ethernet communications via these frequencies which are commonly referred to as 802.11a, 802.11b, 802.11g, and the latest, 802.11n. 802.11n has a throughput of 110 Mbps.

The user's distance from the Wi-Fi antenna, the uses of the same unlicensed frequencies by other devices, and the obstacles within and the structure of buildings which could interfere with the radio signals all affect the communications bandwidth received from the Wi-Fi antenna. Typical coverage areas indoors for omni-directional Wi-Fi antenna are 100 to 300 feet.

In the past, the typical use of Wi-Fi in buildings has coexisted with other wireless systems using the 2.4 GHz radio frequency such as Zigbee and Bluetooth. However, Wi-Fi has started to gradually move into the building control systems. Examples include Wi-Fi temperature sensors (temperature@lert product), Wi-Fi CO_2 sensors (AirTest Technologies), Ethernet field panels with capability to use Wi-Fi transceivers to establish wireless connectivity (Siemens), and Real Time Location Systems (RTLS) (Cisco).

With more sensors, meters, and control devices generally needed in buildings, expect the adoption of wireless buildings systems to accelerate.

4.2.3 Cable Infrastructure

How important is cabling? Cabling, cable pathways, and equipment rooms are long term. Many manufacturers provide 20–25 year warranties. So, what is initially designed and installed is important. Cable may seem mundane but constant advancements in cable throughput, reductions in potential interference and increased low voltage power over the telecom cable requires the installation of the latest cable technology.

A smart building will take advantage of similar cable standards for IT (EIA/TIA 568) and BAS (EIA/TIA 862) and utilize the same cable types (twisted pair copper cable and fiber optic cable). This provides an opportunity to reduce the number of cable contractors, reduce coordination from the construction manager, and share cable pathways. In addition, one cable can supply communications and provide low voltage power to devices on the network via Power over Ethernet (POE). Thus POE eliminates one power cable, and reduces material and installation time and costs.

The smart buildings approach is to maximize the use of the structured cable plant for access control, video surveillance, building automation, and other building systems. A smart building will also supplement the wired infrastructure with the use of wireless systems; deploying Wi-Fi throughout a building, installing a distributed antenna systems (DAS), and using Zigbee, Enocean and RFID in building control systems and asset management.

All of the telecom cabling generally terminates in and is managed from the telecommunications room. These rooms should be served by dual telecommunication entrances, physically separated dual power feeds, and emergency power. The telecommunications rooms need to be remotely monitored for security, water, temperature, and seismic events. The entire infrastructure: cabling, equipment, racks, and wireless access points needs to be labeled with predetermined naming conventions and documented for asset management and daily operations. As IT continues to penetrate our buildings, and our lives, this infrastructure becomes increasingly critical, and a key attribute of a smart building.

How can building owners and designers save money in specifying and installing a building's technology infrastructure? Or more importantly, how can they make sure the money spent is for the best value? Here are a couple tips on saving upfront construction costs and ensuring longer term value for the building.

4.3 Construction Costs

4.3.1 Converge The Cabling Types

There are probably 10–15 different technology systems in a decent size building. The telecom systems have standardized on unshielded copper twisted pair and fiber optic cable. The building automation systems have also standardized on twisted pair and fiber optic cable. While the life safety systems have not standardized, some life safety systems, such as video surveillance and access control systems, are using twisted pair and fiber optic cable, and some cable manufacturers offer full solutions for life safety systems using twisted pair and fiber optic cable. The point is that the more you use standard twisted pair and fiber optic cable, and the more you get away from propriety cabling systems, the greater the opportunity to economize and save money.

4.3.2 Coordinate Pathways for All the Technology Systems

Economize not only on the cable but also the cable pathways. While the cabling of many technology systems have different end-points (such as tenant offices, surveillance cameras, HVAC fans, etc.), the longer backbone runs of cabling through a facility may be common pathways. Install one common highway or pathway for the cabling; try not to build several smaller pathways.

4.3.3   Reduce the Number of Cabling Contractors

Over half of the cost cabling is labor. Why have one contractor install a cable from room A to room B, only to have second contractor install a cable for a different system from room A to B? Why pay twice when the cost for one contractor to install both cables at the same time is marginal?

4.3.4   Use a Client's Master Agreements for the Materials and Equipment

Larger clients, such as corporations with many sites, university systems, school systems, healthcare entities, etc., should have standardized on the materials and equipment they use for their technology systems in their facilities. If they've standardized and are large enough, they should have master agreements with cabling manufacturers or suppliers that an installation contractor may be able to use. The master agreement will have lower prices because of the aggregation of facilities and commitment to suppliers. If you're client does not have master agreements, advise him that he should, and at least make the facility you're involved with the first to avail itself to the lower pricing.

4.3.5   Single Point for Cabling Administration

If you're designing a three story building, do you need to have equipment rooms on every floor? You could serve the entire building from an equipment room on the second floor if the distance between a second floor equipment room and each cabling endpoint is within the distance for standard cabling. There's a multitude of benefits from doing this; less space is utilized for equipment rooms, there's more efficient use of the network equipment, and it's easier from an operational and maintenance standpoint. The small cost for slightly longer cables pales in comparison to the overall savings.

4.4   Operational Costs

The building and the technology infrastructure (at least the cabling, cable pathways and equipment rooms) have long lifecycles, and you'll want to balance the initial cost of the technology infrastructure with the long term value for the building. Here are a couple things to keep in mind:

4.4.1 Warranties

Major cabling manufacturers offer long term warranties on their products and systems. If their products are installed by certified installers and the installation is inspected by a representative of the manufacturer, they will warrant the cable system for 15 to 25 years. Some companies warranty the installation and others warranty both the installation and future technical applications for the cable. Products that provide such warranties may be slightly more expensive, but probably offer better long term value.

4.4.2 Expansion

Many times the size of equipment rooms are questioned—large equipment rooms are provided only to find that a few equipment racks are initially provided. Or cable pathways are provided and only half of the pathway is initially used. No it's not a waste of space. It's long term thinking. You don't ever want to move an equipment room. In addition the emerging building technology systems will demand more space, more power, more air conditioning, and more grounding. Build it right once, and avoid the tremendous disruption later.

4.4.3 Use Cabling Consolidation Points

Cable consolidation points are typically used in open office environments. Instead of running each cable from the equipment room to each modular office or cubicle, a consolidation point is installed close to the cubicles. You may run less cable from the equipment room to the consolidation point, and from the consolidation point you have shorter runs to each of cubicles. The initial cost of the consolidation point is about the same as running all the cabling from the equipment room; the savings is in recabling when the cubicles are moved around, which they typically are over time. Many modular furniture manufacturers have consolidation points hidden in their products that fit in with their modular furniture.

Owners and designers are tasked with controlling initial construction costs while providing for long term value in the buildings that are being designed. It can be quite the balancing act. The technology system design can contribute to that effort.

4.5 Security

The building automation industry is now at a point where we have legitimate and reasonable concerns regarding the security of building control systems, especially in smart buildings where advanced technology is deployed. We see stories in the news regarding malicious cyber-attacks on private companies, government networks and internet sites, and there are questions as to what such an attack would mean for building control systems, building operations, occupants, and owners. The apprehension is amplified in newer buildings because there has been increased penetration of IT infrastructure in building control systems and greater integration and interconnection of building controls with other systems. The potential security vulnerability of a building can extend to the smart grid as we move to implement two-way communication between buildings and the grid, and of course it could also impact corporate business systems. The overarching security concern is more about network security and less about physical security, although the two are certainly related.

For a smart building it is a prerequisite to implement a secured converged network. In addition, the building should have:

▶ An integrated video network

▶ Security Measures—Network admission control

▶ Security Measures—Network intrusion detection

▶ Security Measures—Ability to segment or isolate the network to limit access temporarily or permanently

▶ QoS management

▶ Bandwidth management

▶ Core equipment and cabling redundancy

▶ ISP redundancy

▶ Uninterrupted Power Supply (core network)

▶ IP device management system

▶ Monitoring of energy usage of equipment at the data center/MDF/IDF

▶ Enhanced security elements for integrated networks

▶ Assign an administrator for building control systems with responsibility for ongoing network security.

▶ Utilize IT security measures for the building automation networks.

▶ Provide physical security in areas or spaces where BAS equipment is located

❭ Encrypt and safeguard network traffic.

❭ Secure any wireless network

The threat is that someone can penetrate a building's systems via an un-secured network to cause damage, disruption, theft or possibly loss of life. For traditional IT systems in a building, the threat may be loss of communi-cations, unauthorized access to sensitive data, theft of intellectual property, disruption of equipment (which may include physical security systems such as access control and video surveillance), loss of data, and impediments or stoppage of normal business operations. For the other building systems, such as HVAC control, electrical distribution, lighting, and elevators, the threat is disrupting critical building infrastructure, which also impedes or stops normal operations. Depending on the building use and building con-trol system, a security threat may be related to life safety; for example, dis-rupting emergency power, lighting, and HVAC in a critical healthcare space. The threat to building systems is not hypothetical; the infamous Stuxnet cyber-attack in 2010 eventually affected programmable logic controllers (PLC), a controller heavily used in industry, but also commonly used in buildings, for example in elevators and lighting equipment.

In general the building automation industry and facility management has treated the security of building control networks as a secondary or tertiary issue, if at all. The most popular security approach for a building management system (BMS) is to isolate the BMS; not letting it connect to any other networks. But that in itself is a false sense of security; the BMS

Table 4.1

BAS Security Attacks

Network Attacks		Device Attacks	
Interception	Network Sniffing	**Software**	Code Injection
			Exploiting Algorithm Weakness
			Availability Attacks
			Configuration Mechanism Abuse
Fabrication	Insert Malformed Messages	**Side-**	Time Analysis
	Insert Correct Messages	**Channel**	Power Analysis
	Replay Old Messages		Fault Behavior Analysis
Modification	Man-in-the-Middle Attacks	**Physical**	Eavesdropping
	Alteration		Microprobing
			Component Replacement
Interruption	Denial of Service		
	Network Flooding		
	Redirection		

at a minimum will have fire systems, HVAC, access control, elevators and possibly lighting connected into it, potentially allowing access from one of those networks or one of the devices on those networks. Some minimal or partial security measures may be in place for some buildings, but not the comprehensive security measures needed to prevent or minimize network vulnerability. It's fair to say that most traditional building management systems are not secured.

In fact, many legacy BMS systems have *back doors* allowing the BMS manufacturer or local control contractor to monitor, manage or update the systems. It is interesting that while the recent security concern is about newer intelligent buildings, it is older buildings with legacy BMS systems that are much more vulnerable to attacks. The legacy systems have less computing power and are vulnerable to newer, more powerful and advanced technology that a hacker may use. The legacy systems are also likely to be running older operating systems, some of which may no longer be updated with security patches. In addition, the vulnerabilities of older systems are well known to hackers, thus minimizing the effort and time needed for an attack.

The automation industry has rightfully strived for systems standards with a move from proprietary implementations to open and transparent communication protocols. There are many benefits to open standards: compatibility of products, customization, avoiding being locked-in to one manufacturer, interoperability, competitive costs, and more support options. At the same time, open and transparent standards would seem to increase vulnerability of BAS networks, basically providing all the information hackers would need to assess vulnerabilities and potential approaches for an attack. This may look like something akin to giving the car thief the keys to the car.

But one of the upsides of the open standards movement is that it allows those communication protocol standards to incorporate network security related attributes into the standards. Most major BAS standards have incorporated some security mechanisms into their standards. The security aspects of BACnet are probably the most advanced. But at the other end of the spectrum is Modbus, which has no inherent security capabilities.

A cyber-attack on a BAS network is either going to access the network, trying to access or disrupt the communication or exchange of data, or the BAS devices, namely the controllers, actuators, and sensors. The BAS network could be accessed physically, or possibly via wireless communication, but also through a network device, such as a compromised controller. The attacks on the devices are likely to come from the network or physical manipulation of the device.

Table 4.2

Typical IT Security Measures

Strong firewalls
User authentication
Secured wireless
Awareness about physical security
Use VPNs in enterprise situations
Back-up policy
Strong encryption of BAS data communications
Network hardware is in secured data center
Intrusion detection systems
Devices that can capture IP packets

4.5.1 Tips on Preventing a Security Breach

Developing, testing, and deploying security measures in buildings needs to be an ongoing activity, built into the operation of the building. Here are some suggestions for the first steps:

Assign a dedicated network administrator for building control systems, with the responsibility for ongoing network security. The network administrator should coordinate security efforts and responses, as well as internal and external assistance.

Baseline your network. Know what normal traffic is. Identify the indicators of an attack. If facility management is spearheading the effort, bring in your IT department early on.

Take a comprehensive approach. Assess every building system, the vulnerabilities, and what the loss or disruption of the systems will mean to building operations and occupants along with the financial impact.

Start with the use of IT security measures on the building automation networks. Understand that while the IT security measures are valuable they may not apply to all systems or portions of building control systems. For example, at the field or application control level, you may find controllers with limited processing power and memory utilizing a limited bandwidth network.

1. Provide physical security in areas or spaces where BAS network cable runs.

2. Encrypt your network traffic.

3. Secure any wireless network

Take into consideration the human aspects of security; the greatest threat is from the inside; disgruntled employees, employees taking shortcuts or bringing in their own laptop. Develop policies on passwords, configurations, settings, and a training program.

Comprehensively securing a building not only involves access control and video surveillance, or an IT security program, but must also include the building control and automation systems.

4.6   Communication and Data Infrastructure

The method of communication and data exchange within and between building systems is vital. It is a foundation that will determine the difficulty or ease of integrating system functionality and data. Smart buildings shun proprietary protocols in favor of standard open communications protocols based on the ASHRAE BACnet I/P, OPC DA, Modbus TCP, oBIX, XML, SOAP and SNMP standards of data exchange or similar open standard protocols. Many building products now incorporate open protocols, some going through a process that verifies or certifies their adherence to the protocol standard.

The network architecture of building systems should take into account the minimum speeds for serial buses, maximum size in points and devices per serial bus, maximum number of serial buses per network controller, and the use of native open protocol controllers versus gateways in existing buildings.

The adherence to standard open protocols and detailed network design extends not only to building control systems but also to facility management systems, business systems, and IT systems in the use of standard database structures such as SQL or ODBC, Oracle, or DB2.

4.7   Facility Management Software

The operation of a facility represents the longest duration of the lifecycle of a building and the largest portion of the total lifecycle cost. The facility management responsibilities are wide-ranging and multidisciplinary; varying from responses to immediate alarms or emergencies to long term capital planning. Throw in the fact that today's buildings are increasing more complex, and the need for a smart building to have top of the line facility management tools and processes to effectively manage the facility becomes essential. The software facility management applications shape the administrative processes within facility management. These systems are discussed in the following sections.

4.7.1   Work Order System

This is a system that can initiate work orders, assign tasks to internal staff or third party contractors, track the work orders, and archive records. The system should allow for analysis of work orders based on building, space, type of asset, personnel. The system should also track labor, materials, travel expenses and assign relevant work orders.

4.7.2   Preventative and Predictive Maintenance

This application is used to prevent failure of equipment through the use of suggested preventative maintenance from the manufacturer or by predicting equipment failure based on equipment data. The system should support the scheduling and tracking of recurring maintenance tasks, automatically assign tasks and work orders, and be able to create daily, weekly, monthly, and yearly tasks. Typically this maintenance may involve inspections, testing, measurements, and parts replacement or adjustment.

4.7.3   Space Planning

This is a software application allowing effective space allocation for the use of the building by tenants or owners. It provides the capability to draw and manage space with two dimensional floor plans using actual sizes of real world objects such as furniture and equipment, walls, windows, doors, etc. It allows for the development of alternative space layout and provides the basis for real and well-organized space design and planning.

4.7.4   Material and Equipment Parts Inventory Control

This is basically a system that tracks inventory and identifies inventory needs, possibly automatically triggering purchase orders of parts and equipment based on certain thresholds.

4.7.5   Asset Management

This is a critical application that should house all the details of the facility assets. It should contain all the details for each piece of equipment. This should include the name of the item, identifying serial numbers, location, warranties, manufacturer and its maintenance history. The asset management application must be integrated into the preventive maintenance application and provide data on when preventative maintenance has to occur and generate the work order.

4.7.6   Data standards

Data is an asset and facility management should have written standard methodologies and processes to manage the facility data. This would include document management, naming conventions and standardized databases, in coordination with other relevant groups and applications, such as the IT department and enterprise asset management.

4.7.7   BIM Integration

Building Information Modeling is the significant data management tool for new construction. The data and modeling generated in the use of BIM must be exported in the facility management system (specifically the COBie files into the asset management application.) This export of design and construction data into the facility management system is critical to the successful handoff from construction to operations. Without it FM and the building operations are handicapped at the outset.

CHAPTER

5

Contents

The Management of Building System Data

A wealth of information creates a poverty of attention
—Herbert Simon, Economist, 1971

5.1 Overview

There's a lot of interest in the industry today about dashboards, especially energy-related dashboards. While dashboards are ultimately the user interface into building and energy data, very little attention has been given to the quality and management of the data behind the dashboards. Dashboards are somewhat like the cover of a book. The data is the book and the fundamental underpinning of the information provided to the user who's looking at the cover. The best designed most intuitive dashboards are useless if the dashboards utilize inaccurate or incomplete data.

Building system data must be viewed as an asset: it has value, is necessary for properly operating and maintaining the building, and it must be managed and treated as such. The question is how to get accurate, validated, and well organized data from our building systems that can be managed on an ongoing basis. What follows are some of the issues we face in managing building systems data.

5.2   Lack of Planning

Most building operations do not have a data management plan. What passes for the data plan consists of a database associated with their Building Management System (BMS). With this approach the information is limited to those systems monitored or managed by BMS. In addition, the setup of the database, the naming conventions format and structure is left to the BMS contractor, usually with little or no input from the building owner. Furthermore, getting the data out of the BMS database often involves additional software or application programming interfaces (API) from the manufacturer, sometimes even multiple copies of the same software or APIs based on data points limitations of the software or API.

Comprehensive planning means that one should take a broader look at all the data and information required in order to manage the building's performance. The data management plan needs to be put in writing and a standard operations policy—no more makeshift, improvised approaches. An investment of time at the start to properly develop a data plan will save time later.

Start the plan with a wide-ranging scope. Identify the data and information that different people need to perform their work. Much of the data will be monitoring points on building systems, but, some data may be needed from business systems or other systems outside of facility management or even outside the organization.

Identify where the data exists or how it will be generated and collected, and estimate the scale or volume of data. Decide on a data format. Deal with the administrative aspects of the plan such as user access, dissemination of the data, how data will be integrated, how it will be archived, retention of data policies, and how often the plan is reviewed. Plan the organization of the data to assure the data is accurate and easily accessible.

5.3   Standardized Naming Conventions

Standardize what you call things. A multi-building campus with buildings built at different times with different contractors is likely to have multiple names and tags for similar pieces of equipment. You don't want to end up with ten different names for air handlers or pumps. Multiple naming conventions in an existing building or portfolio of existing buildings is the largest and most time consuming issue involved with implementing an integrated building management system.

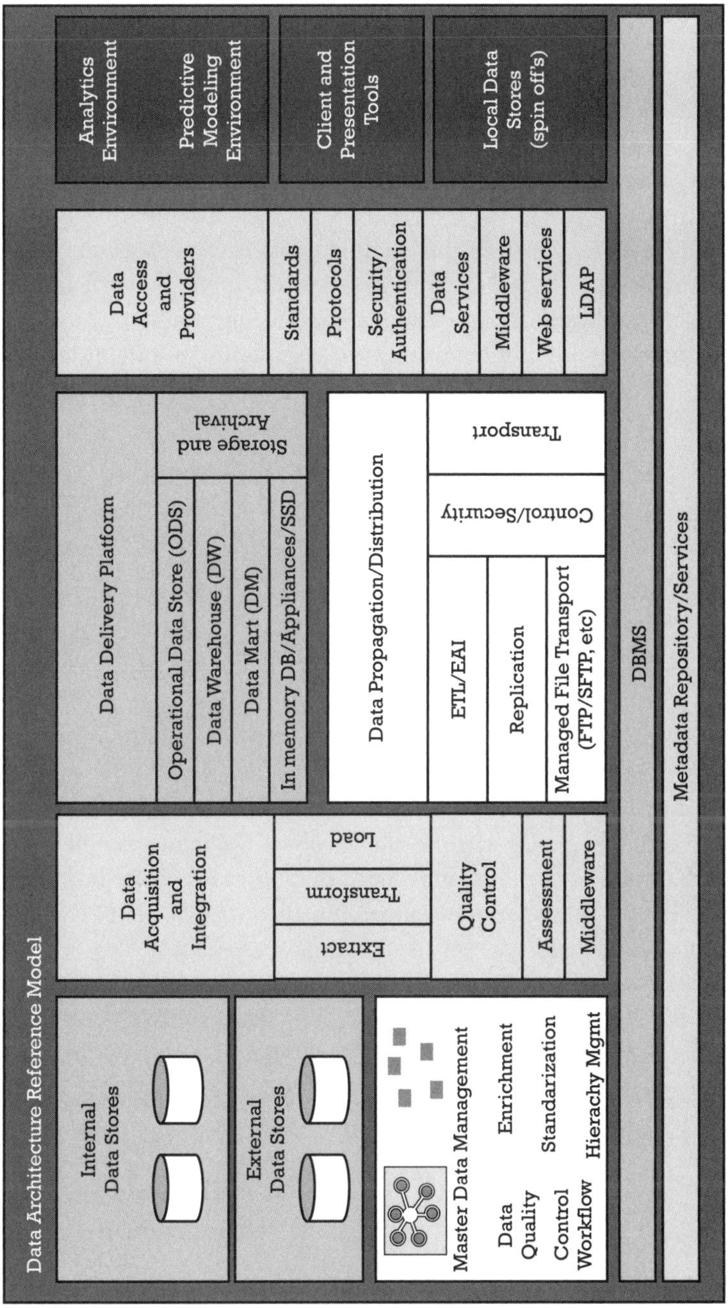
Data Architecture Reference Model
Internal Data Stores
External Data Stores
Master Data Management
Data Quality
Enrichment
Control
Standarization
Workflow
Hierachy Mgmt
Data Acquisition and Integration
Extract
Transform
Load
Quality Control
Assessment
Middleware
Data Delivery Platform
Operational Data Store (ODS)
Data Warehouse (DW)
Data Mart (DM)
In memory DB/Appliances/SSD
Storage and Archival
Data Propagation/Distribution
ETL/EAI
Replication
Managed File Transport (FTP/SFTP, etc)
Control/Security
Transport
Data Access and Providers
Standards
Protocols
Security/ Authentication
Data Services
Middleware
Web services
LDAP
Analytics Environment
Predictive Modeling Environment
Client and Presentation Tools
Local Data Stores (spin off's)
DBMS
Metadata Repository/Services

Figure 5.1

Building system data is no longer only for facility management and building operation, but, also has value for larger business processes such as asset management, capital planning, regulatory compliance, energy management and a host of other business applications. This basic reality cries out for a comprehensive information and data management plan for a building or a portfolio of buildings.

Data does not belong to just one department. If you take that approach you end up with silos of data and miss opportunities to create meaningful metrics and key performance indicators. Consider coordinating the naming convention or a portion of the naming convention across systems (i.e. BMS, asset management, purchasing, work orders, etc.) so that other departments and applications can understand and share the data.

The format of a naming convention for data and equipment is less important than strict adherence to and enforcement of one standard naming convention.

5.4   Data Mining

Facility managers are missing opportunities if they don't have the analytic tools to mine, predict, and correlate building data. Other organizations and businesses mine data from their users and customers and analyze the data in order to predict and guide their business. Data mining has been around for a while and is used extensively in web sites, retail purchases, financing, and smartphones, to name a few. A retailer like Wal-Mart knows how many rolls of paper towels are sold daily at each store location, data that is part of a process to optimize their just-in-time supply chain process. Yet, how many large building owners can even tell you how many people entered their building on a daily basis or which building space is the least energy efficient? Which is the most used space? Which is the least used?

Data mining related to energy usage would seem to be a wide open field. Energy consumption metrics related to space usage, operations processes, and business aspects can provide new insights. Predictive value means the organization can be proactive rather than reactive.

5.5   Validation of Data

There's no point in collecting inaccurate data. To get the most accurate information you'll need to tune-up the building systems and check the calibration of sensors and meters. The building systems themselves should be regularly recommissioned or better yet continuously commissioned using a real-time building system analytic tool. Traditional commissioning uses the design documents and design intent for the foundation of commissioning.

Over time however, building spaces or uses may change. So that while you can confirm or validate the design parameters, for example 54° air being delivered by an air handler, the space may have changed and have a different cooling load, not needing 54° air from the air handler. There the systems may need to be adjusted in order to reflect current conditions.

Sensors and meters should be regularly calibrated, including the device itself as well as the communication between the device and its controller. Inaccurate sensors may provide a false sense of complacency and more importantly waste energy and money. For example, assume you have a temperature sensor that is 2°off, showing a discharge air temperature of 55° when it's actually 53°; this two degrees may trigger extra cooling and additional power consumption by the chiller and air handler or reheating of overcooled discharge air which obviously wastes energy.

5.6 Document Management

How much time do we spend trying to find as-built drawings or some similar dated documents? While the building systems' data points can be part of a typical database, a significant portion of relevant facility management information is likely to be in other formats. These formats include paper, such as hard-copy drawings, submittals, O&M manuals, photographs, contracts, faxes, and forms, but, also electronic files in Word, PDF, Excel and Autodesk, all of which need to be managed. A document management system should be implemented to scan the paper documents into an electronic format and store all of the electronic files.

The system typically has an index with a format that may be similar to a building database, which either stores the document or directs users to another system where the document resides. Systems typically have a search capability allowing users to retrieve documents based upon different criteria. These systems compliment the data management plan and in order to truly integrate they need intuitive indexing with firm adherence to the administrative processes of indexing and document conversion.

With the flood of data potentially generated in a building it's not unusual or unexpected to feel overwhelmed. With the proper administrative processes the technology generating the data can be put to work tofind the valuable information we need to effectively manage our buildings.

5.7 Benefits of Data Management

There would seem to be a very good case for bringing all the facility data into one unified database architecture and putting into practice standard methodologies and processes to manage the data. There are several benefits to this approach:

▶ Building data would be more widely available and sharable: Setting aside confidential data, more data would allow for additional analytics, possibe new correlations, metrics and insights into the building's performance.

▶ Building data would be more easily accessible: Have you ever looked for as-built drawings or equipment spec sheets, only to discover that they are not where they should be? Without a structured approach to data management you waste time internally because of the disorganization in the data and documents. Many times building operators will need to contact the original architects, engineers, or contractors for data, thus wasting time and money. What's needed is an orderly index as part of a larger data management system. A structured approach to indexing is vital as facility data grows, which is obviously very likely.

▶ A structured approach can improve the archiving, preservation and retention of data for the long-term: There's some data and information you'll want for the life cycle of the building and there are analytic opportunities in long-term data you'll want for comparison and trending.

▶ A comprehensive data management plan would improve the integrity of the data: Bad data is worthless data. You want accurate, reliable, consistent, and complete data. A structured approach initially validates the data, and then puts into place a process where the data can't be changed or destroyed without authorization.

▶ Streamlining data: There are roughly 6,500 languages spoken in the world today; for data management, you only want one language of standard naming conventions, formats, indexing, and data descriptors. This makes it easier to access and understand the data.

▶ Improving efficiency: We don't organize data for its own sake, but, do so in order to maximize the effectiveness and efficiency of operating buildings. A structured approach can provide additional opportunities for greater correlation between data, improved data analytics, and the possibility of developing or identifying new building data metrics.

5.8   Practical Data Management Activities

Programming: If you're involved with new construction, going through the programming and conceptual design of the facility, the project team should identify a person or group that establishes rules for the data management of data generated throughout the project. While the focus in new construction is typically the construction schedule and the capital budget, acknowledgment and appreciation of long-term operations, rules and standards for data management, and the need to exported data into operations and facility management systems is positive.

Building Information Modeling: BIM is the significant data management tool for new construction. Data can be generated and stored in the BIM CO-Bie files throughout the process of design, construction, and commissioning. The updating of data occurs several times during the project. Responsibility for the data is shared and shifts from the designers to the contractors during the project. BIM data also needs to be updated based on RFIs, construction related changes, and change orders.

Submittals: Construction submittals are an important milestone in new construction or renovation. Submittals usually involve shop drawings, product data, samples, and coordination drawings. Quality assurance and quality control submittals involve design data, test reports, certificates, and manufacturer's instructions. Submittals must be in an electronic format; all of this data and information needs to be provided in an electronic form, preferable COBie for the product data, or a format that is part of a building owner's data management system.

Systems Integration: We generally integrate buildings systems to enhance functionality; integrating fire systems, access control system, elevators, and HVAC are the best examples. We also integrate systems when building owners have multiple BMS systems but want one overall enterprise platform. In that case, the larger integration platform acquires data from multiple systems in various formats using different communications protocols. Through the use of middleware the data is standardized and one database is created, much like a data management system may use. In some cases the standardization of data to facilitate advanced building management systems is in alignment and could be used with an enterprise data management system.

Commissioning: During commissioning and project closeout, data and information such as commissioning reports, project record documents, contract drawings, project manuals, contract modifications, startup logs, test reports, certifications, and the complete as-built BIM is generated. All this information should be permanently retained and accessible. Some documents may be paper, such as certifications, but all documents and data should be submitted electronically and stored. If the building or systems are modified, the designers and contractors will want to use the original record document as the baseline.

An immense volume of building data is created during the design, construction and operation of a facility, but, we've only managed and analyzed a relatively small amount of the available data. The industry foray into data management and analytics is just in its infancy. The initial results, however, especially FDD applications, are impressive. We should expect the FDD model and other analytic methods to apply to other building systems and additional data to be generated by new building systems, such as indoor positioning systems, motorized shading and water reclamation. The starting point is a facility manager given the responsibility for implementing a structured data management system.

5.8.1   The Role of a Facility Data Manager

If you think data is an asset, then you need a person to manage the asset.. During design and construction of a building, data will be generated. It is in the operations of the building that data not only will be generated, but, also consumed. Given that building operations and maintenance is the most expensive part of total life cycle costs, and the longest time duration within the building's life cycle, we need data management during every building phase: design, construction, and operations.

During design and construction, we typically have two to three people tasked with managing various data. One is the LEED consultant tasked with gathering energy and sustainability information for the building certification. Another is the BIM consultant organizing BIM models and data. The third is the architect who uses project management software to communicate and share data with the project team. After commissioning or occupancy of the new building, the roles of the BIM and LEED consultants, and possibly the architect, expire.

The facility manager tasked with data management would have a responsibility in implementing the data management system for the building along with the acquisition and management of the data from the initial building design through construction and facility management. Facility managers design, deploy, maintain, monitor and even enforce a comprehensive program for data management.

5.9   Dashboards: Transforming Data into Information

Data is raw material. Its real value is in transforming the data into useful information. Information is the finished material when some intelligence has been gleaned from analyzing or studying the raw data.

Building systems can provide a lot of data through BAS points, sensors, meters and measurements, yet yield little information. Many of the graphics used to display data in a traditional BAS system seem like legacies.

They require enormous setup time and are generally reused from job to job. Today's more advanced integrated building management systems use browser-based dashboards for the human-machine or user interface.

It is digital dashboards that can provide relevant and timely information to several levels or groups involved with a building's performance. These different users can be facility technicians, managers, C-level executives, tenants, occupants, or visitors via kiosks or a web page. The information provided may cover the specifics of particular building systems such as HVAC, electrical, or specialty systems, but, tend to focus on energy usage, costs, performance trends, alarm management, comparisons with similar buildings, and KPIs.

The key to the use of dashboards is twofold. First, one needs to determine the right information for the viewer of the dashboard. Facility technicians have different information needs that C-level executives, though some of the data to create the information may be similar. The second key is the creation of the user interface and interaction with the dashboard, which involves visual design and human factors. This is more about how humans perceive, evaluate, process, and act on information. What follows are some tips on creating the most effective dashboards.

Information can be emphasized or de-emphasized by its position on a dashboard. The visual dominance is the in the center of the screen. Depending on how the culture reads (left to right, or right to left) the other area of dominance will either be the top left or top right of the screen. The other corners are neutral, or in the case of the bottom right, actually de-emphasized. So the most important data should go in the center or the top left of the dashboard. This is especially true if other secondary data on the dashboard can only be understood based on understanding the primary information.

5.9.1   Facilitate Comparative Analysis

When you have two charts that are meant to be contrasted, it is best to arrange them side by side. This arrangement signifies the need for comparison.

When you have a very narrow range of data, scale the chart to make the data fill the chart. This makes it easier for the user to analyze the data.

5.9.2   Customize Chart Scale For Optimal Data Presentation

Sometimes data displayed on a chart has a very narrow range. This makes the task of analyzing the data more difficult. Such situations call for manipulation of the chart scale. The chart scale should be adjusted so that, its lower and upper limit are close to the lower and upper limit of the data range. This adjustment will help in accentuating the ups and downs of the plotted data, thereby making analysis easier.

5.9.3   Appropriate Selection of Charts

For maximum impact, it is essential that you choose the right chart for you data. The pie chart is often used inappropriately. A pie chart is actually meant for plotting percentages, but, it is sometimes used for plotting non-percentage data such as sales, revenue, and quantity.

5.9.4   Proper Formatting of Numbers

It doesn't make sense to have a chart that displays numbers with excessive accuracy. Neither does it help if the chart is cluttered with very large numbers. So, it is best to restrict the number of decimal places to 1 or 2. And, scale large numbers by defining a proper scaling parameter. The K,M scale can be applied to financial charts to scale down numbers that are greater than thousand and million.

Dashboards are an important aspect of business management systems and are referred to during the planning and decision making process. Therefore, it is essential to make the dashboard uncluttered and user friendly.

Dashboards have become popular in recent years as uniquely powerful tools for communicating important information at a glance. The greatest display technology in the world won't help if you fail to use effective visual design. Although dashboards are powerful, their potential is rarely realized. And if a dashboard fails to tell you what you need to know in an instant, you'll never use it, even if it's filled with cute gauges, meters, and traffic lights. Don't let your investment in dashboard technology go to waste.

This book will teach you the visual design skills you need to create dashboards that communicate clearly, rapidly, and compellingly. Information Dashboard Design will explain how to:

▶ Avoid the thirteen mistakes common to dashboard design;

▶ Provide viewers with the information they need quickly and clearly;

▶ Apply what we now know about visual perception to the visual presentation of information;

▶ Minimize distractions, cliches, and unnecessary embellishments that create confusion;

▶ Organize business information to support meaning and usability;

▶ Create an aesthetically pleasing viewing experience;

▶ Maintain consistency of design to provide accurate interpretation;

▶ Optimize the power of dashboard technology by pairing it with visual effectiveness.

Designing an effective UI for a dashboard can seem a deceptively simple task at first, but soon presents challenges sufficient to ensnare even the most deliberate of developers. The fundamental challenge with dashboard visualization is the conflict of having a limited amount of space in which to communicate a lot of information

Ideally dashboards are more than just business scorecards. They should do more than attempt to balance financial success with perceived business processes, in order to generate growth. Dashboards should provide users with the ability to relate and analyze large patterns of information, in order to glean new insights or understand the business condition more fully.

The UI should be designed to support the specific types of business decisions and insights users hope to gain by viewing the dashboard. This clearly requires more than just a basic understanding of the user's intent, lest we end up with a potpourri of gauges, charts, and tables that only superficially describe the health of the business and fail to uncover any meaningful or directed insights. Providing the user with an effective and engaging experience requires the effective use of layout, color, and interactivity in a synchronistic manner.

But the challenge with dashboards is even bigger yet. Beyond visualization and UI lie a unique set of engineering requirements for summarizing, aggregating, and pre-calculating information computed from large data sets. For instance, showing basic sales figures during a period of time filtered by product SKU would most likely entail gathering source information from some type of ERP transaction system on an order or line-item basis. A company with any significant order volume could easily be referencing millions of transactional records which would need to be summarized across product SKU.

Adding other filtering dimensions like geography, channel, market segment, etc., quickly creates need for some type of back-end data engine to aggregate and pre-calculate this summary information. In most cases these back end data repositories take the form of a data warehouse, with some type of online analytical processing (OLAP) system to process the data within the warehouse.

Yet for many organizations, the data warehouse will never encompass all the information that needs to be referenced. Or perhaps they may not have sufficient business intelligence systems in place to collect and aggregate the information. In these cases, calculations tend to be performed by the most pervasive analysis tool employed by businesses today - the spreadsheet.

Business analysts collect data from several sources by hand in the form of extract reports, then import it into a spreadsheet and perform manual calculations to prepare reports for consumption by higher level executives. This manual aggregation process occurs at varied intervals depending on the demands of the business and may work in isolation or in conjunction with more automated back end data aggregation.

In both of these scenarios (using a data warehouse/OLAP solution to pre-calculate or manually aggregating and calculating data via spread-sheets) the dashboard designer is faced with the same data structure challenges. The dashboard will need to collect data from several sources at once and present that data in some type of hierarchal format representing different levels of aggregations and calculations. Most dashboards need some way to transform, filter, and drill across hierarchal data structures, making it critically important to engineer a solution which accommodates these structures in a flexible and consistent manner.

When it comes to a development methodology for building dashboards there is no silver bullet solution. However, there are some general principles which have been found to be effective.

5.9.5 Prioritizing Users Over Data

First among these principles is the choice of basic orientation. There are two alternate approaches that can be used when gathering requirements for a dashboard project: the traditional "bottom-up" (aka the "data-centric") approach; or the more contemporary "top-down" (aka "user-centric") approach.

The data-centric approach is employed most often by teams with more of an engineering focus. It concentrates first and foremost on providing flexible ways to access the data available to users.

Beginning with the underlying data models, the bottom-up method envisions the user experience through the lens of the data structures and defines the navigation and access points accordingly. Highly efficient in its multi-level data access, the risk in this approach (and the reason many good dashboard designers avoid it) is that it often compromises the subtle requirements of the users in favor of accommodating the data. This may lead to a poor outcome as success in dashboard design is not necessarily defined by having the most flexible access to the widest possible data sets, but rather by offering users quick access to the data most relevant to their specific business needs.

The user-centric approach, in contrast, begins with the questions of "what do we need to communicate, to whom, and to what degree of depth? Grounded in the requirements of the users, this path ensures that dashboard design follows a more intuition-guided experience to maximize the level of engagement amongst the executive constituents.

While this approach may seem obvious to those with expertise designing UI for web site applications, many engineers more accustomed to working on back office functionality (e.g. database and report design) might not be as familiar. Rest assured that data structures in this approach are no less important, but often need to be bent and molded through filtering and summarizing techniques, which can result in table structures that are inefficient in the classical data sense, but optimized for a superior user experience.

Traditionalists often raise concerns about minimizing the importance of sound data management principles, in particular when large volumes of transaction-level data are involved. However, most dashboards are not used as data-mining tools seeking insights at the lowest level of data capture, but rather performance summary systems offering a broad view of things with one or two levels of drill-in capability to answer the most common questions.

The user-centric approach requires a keen understanding of the user's needs, and the ability to design around them. This can be reduced to two fundamental questions:

What decisions or insights does the user hope to make by interacting with the dashboard?

What data is needed and how does it need to be presented to answer those questions?

In some cases, after gathering user requirements, it becomes clear that key pieces of data needed are not stored in the current data repository. In order to add meaningful insight to the data already available, it becomes necessary to devise a way to incorporate the missing elements. In a bottom up approach, this realization may not have occurred until the dashboard was deployed to the user audience.

Trying to design an OLAP solution for a dashboard without clear requirements from end-users is a dicey proposition, mostly because users often have a difficult time articulating requirements until they've had an opportunity to tactilely experience the dashboard for themselves by pointing, clicking, dragging, and dropping. Prior to having a chance to "play" with the UI, users struggle with the prospect of validating back end data schemas and data structure requirements as these artifacts are too far removed from the real world use of the dashboards.

5.9.6 The Benefits of Management Dashboards

▶ Fully customisable, provides users with all the analytical information necessary to efficiently fulfill their roles.

▶ It provides a clear view of company KPIs to enable analysis of defined objectives.

▶ Intuitive, easy to use with rapid user adoption.

▶ Flexible, tailored to meet the specific goals, objectives and strategies of your company.

▶ Measure performance of departments.

▶ Measure Your Return On Sales & Marketing.

▶ Provides right information to the right people at the right time to optimise and accelerate decision processes.

Executive Dashboard project participants are fortunate when they have the opportunity to work with true graphic designers on the visual design portion of the dashboard. There is a real opportunity to see that design is a discipline and that the value the visual design team is not merely skin deep, but integral to the success of the dashboard interface. Of course, graphic designers vary in skill and understanding, but you know you are in the presense of a real master when they teach you the fundamentals of their trade. When they expose the "why" of their decision making, you learn that making your business intelligence dashboard intuitive relies on understanding factors of human perception.

Let me illustrate the point with a great example from the incredible book, *Designing Interfaces* by Jenifer Tidwell. In the chapter on "The Basics of Information Graphics," Tidewell points out a topic of great interest and value: Pre-Attentive Variables.

Basically, the idea is that you have to choose your visual features very carefully because they operate "preattentively"—that is, they convey information before the user pays conscious attention.

Now that you know how important it is to use preattentive variables when you need to impress upon your user a quick understanding, wouldn't you like to know what the actual preattentive variables are? Of course, so here you go. Use these techniques in your business intelligence dashboards and reports and enjoy the level of understanding they can bring:

- Color
- Position
- Brightness
- Orientation
- Saturation
- Size
- Texture
- Shape

5.10   The Handoff Between a Newly Constructed Building and Building Operations; How Not To Fumble

Example is not the main thing in influencing others, it's the only thing.
—Albert Schweitzer

One reason a building may not perform as well as it should is often related to how the newly constructed structure was "handed off" to building operations. A poor transition process may mean the building operations get off to a bad start and never fully recover or only catches-up after much effort. Design and construction phases for a new building have structured processes, and the handoff activities from new construction to operations are addressed in the project specifications. Despite the clear requirements and acknowledgement of close out activities, the transition or handover is often undervalued, misunderstood and overlooked.

Many of the critical elements of the "handover" pertain to the building systems and the relevant data or information regarding the design and construction of the building. However, these are not the only concerns and activities in the move to building operation. Prior to the turnover the building owner will need to determine staffing for maintenance and operations, and then assign and train the staff. Prior to occupancy the owner will also oftentimes be involved with the furniture and equipment, warranties, correction periods, record documents, spare parts, extra materials and specialized operating tools.

Addressing this transition from construction to operations, specifically regarding the building systems and data will improve the initial and long term operational performance in turn saving money and increasing the value of the building. In the following section are some tips on how to avoid fumbling the ball.

5.10.1 Give Operations Personnel a Seat at the Table with the Design and Construction Teams

This happens frequently but not all the time. As active participants the operational personnel is able to provide perspective into the operational and maintenance aspects of materials or design and construction alternatives. This must be tempered with an awareness of the natural inclination to simply operate the new building the same as the old building, especially an issue when the operations of the existing building are sub-optimal. This interaction helps to assure the facility manager is familiar with the new facility and allows the FM an opportunity to develop its O&M program.

5.10.2 Install Some of the Facility Management Software Applications Relatively Early in the Construction Process

Don't do the typical thing and wait until the end of construction. Using software tools, such as system analytics, will assist in commissioning and start-up and installing them earlier in the process facilitates a longer time period for testing the tools and staff training. This initiative is consistent with a larger issue in that technology and specifically some form of an IT network should be in a building under construction earlier as well. For example, if you're installing BAS network controllers that directly connect to an IT network, the IT network has to be installed or you can't fully test the system. Some of the IT components (servers, firewalls, etc.) may be in the "cloud" while the building is being constructed, and then later brought into the building.

5.10.3 Have the General Contractor or Sub-Contractors Operate the Building For a Short Time, and Then Transfer Operations to the Owner

While this model has generally been used on larger infrastructure projects the beauty of this approach is the incentives it provides the construction contractor to get the building operations right. The contractor has "skin in

Table 5.1
Specifications Sections Addressing Project Handoff

Startup procedures
Closeout procedures
Closeout Requirements
Closeout Submittals
Maintenance contracts
Operation and Maintenance Data
Operation Data
Maintenance Data
Preventative Maintenance Instructions
Project Record Documents
Sustainable Design Closeout Documentation
Life Cycle Activities
Commissioning
Facility Operation
Facility Operation Procedures
Facility Maintenance
Facility Maintenance Procedures

the game". While this approach still involves an eventual handoff to the owner, the likelihood of a "cleaner" handoff is often much improved.

5.10.4   Insist On the Use of BIM During Design and Construction

Generally the transfer of data from the design and construction processes into a facility management system is inefficient and ineffective. Most of it is paper handed over in three-ring binders and boxes, supplemented with CDs of drawings and specifications. Lost in this handoff is data and information that could improve the management and operation of the building. Building Information Modeling (BIM) allows for the design and construction data and information to be transferred electronically.

5.10.5   The Most Value That Operational Personnel Can Bring to the Table Is Their Involvement In Defining the Requirements of Commissioning, System Start-Up, and Close Out Procedures

The owner's facility staff doesn't replace the project's commissioning agent or take on the responsibilities of the contractor. However, it is the commissioning, system start-up, demonstration of equipment operation and load testing that are at the heart of close-out activities and the turning over of the building systems. The facility staff's involvement in these processes is essential to a smooth transition.

5.10.6 Identify the Data, Information and Resource Materials Needed to Operate the Building

Operating and maintaining a new facility requires data and information. The building owner, with the participation of the design and construction teams, needs to define the data and information required for efficient and effective operation of the systems and related services during the facility design process. These requirements will be part of the commissioning and close-out activities as well. Typically a contractor provides the owner record documents. These are O&M manuals, record submittals, shop drawings, and record specifications and drawings. Whatever the requirements, they need to be part of the contract documents for construction.

The discussion of data and information needs may also spur interest in more data points, such as meters or sensors that gather data. The process can also identify the need to integrate data into the owner's business systems. If it does, it's a good thing as it demonstrates the project team is looking forward to operational needs.

5.10.7 The Expectations of Contractor's Requirements Must Change From Just Installing Equipment to Completing and Leaving Their Work In a Condition for Long Term Operations and Support

Too many times contractors will install equipment, complete their deliverables and then reduce staffing and move on to the next job prior to completion of close-out activities. The result may be incomplete work or at the least, less than optimal involvement in close-out activities. Contract requirements and the mentality of the project team needs to be focused on the most important and costly part of the building's lifecycle: long term operations and maintenance.

5.10.8 Conduct a Review of the Transition to Operations and Document Lessons Learn

Regardless of your expertise or experience there is always an opportunity to learn from each project closeout or at least confirm that existing procedures are producing the expected results.

During the design and construction processes the focus is often on schedules and budgets. And with that process taking a couple years, building operations may seem far off and something that can be addressed close to completion. Better "handoffs" and transitions prescribe that we embed operations and maintenance into every aspect of design and construction.

Contents

Lighting

6.1 Overview

Lighting affects several important aspects of a building. There is the aesthetic impact it will have on the building and its spaces, resulting in how occupants will perceive and react to the built environment. There is the more functional aspect of providing lighting to facilitate life safety and the work occupants undertake in and around the building. And finally, lighting is a significant consumer of energy. It is estimated that lighting accounts for 21% of electric power consumption in commercial buildings and about 14% of electricity usage and costs in residential buildings. Unneeded and uncontrolled lighting within a building wastes energy and increases facility operational costs.

Different building uses will have different focuses for the building's lighting. These might be as part of building security, to maximize occupant productivity, energy conservation, or just to enhance the emotional reaction of occupants, for example, in retail stores or theaters.

A smart building entails a centralized lighting control system in all the building's usable square footage. The system should provide a dimmable lighting strategy. It also should have a global and zone light schedule and override capabilities. Lighting can also affect other technology systems, such as the need to cool spaces where lighting has increased the temperature. Lighting control systems provide lighting for occupants of the building as needed, but do it in an efficient manner, consistent with any applicable building needs and energy codes.

The requirement for lighting in a building varies by building type, spaces within the building, time of day, and occupancy of the building. Consequently, the control strategies and functions of a lighting control system reflect these variances and primarily involve:

▶ Scheduling: A control system may have a predetermined schedule of when lights are turned on and off.

▶ Occupancy Sensors: For building spaces where occupancy is difficult to predict (such as meeting rooms or restrooms), lights may be turned on and off based on a lighting control system device sensing occupancy.

▶ Daylight: To reduce the need and cost of lighting spaces, a control system utilizes natural light as much as possible. This is sometimes called daylight harvesting or daylighting.

▶ Window coatings: Spectrally selective window coatings, designed for hot climates with large amounts of solar radiation, work by selectively filtering out frequencies of light that produce heat while minimizing the loss of visible light transmission.

The lighting control system distributes power to the available lighting units in a typical fashion, but inserts digital control and intelligence in many of the control devices, such as the circuit breaker panel, wall switches,

Figure 6.1 Lighting in a building.

photocells, occupancy sensors, backup power, and lighting fixtures. The control system significantly increases the functionality and flexibility of the lighting system by providing digital control and intelligence to the end devices. For example, a reconfiguration of lighting zones is accomplished through software rather than the physical recabling. In addition, intelligent end devices allow more focused application of lighting control strategies to specific spaces within the building.

6.2   System Control

The heart of the lighting control center is typically a web-enabled server,, a workstation with a GUI interface, and client software applications for system administration. The lighting control system may be interconnected to other facility technology systems, such as shading or the HVAC system. The networked system allows any authorized individual, including tenants or other occupants, to adjust their lighting through the network or a web browser.

One approach to the lighting control system is the use of intelligent controllers These controllers are distributed throughout a facility, house system electronics, and manage downstream relay panels. The controllers and the system server are networked via an Ethernet network, usually sharing schedules and overrides. The controller may have a user interface panel, which can be used instead of a system workstation, to program and monitor the lighting control system.

System controllers may be modular to allow for growth. The controller may also have several communications interfaces such as an Ethernet port, and ports for RS-232 and RS-485 communications. The system controller communicates with each of the panels through an Ethernet connection, or a BACnet, LonTalk or Modbus protocol that is routed to a larger IP network.

Another emerging networking strategy for lighting control systems is distribution of the intelligence and control further downstream to each device by providing a network interface for each device, such as a lighting ballast. This approach centralizes the control to the network server and allows for network interfaces to specific devices.

6.2.1   Relay Panels

Relay panels are typically mounted next to the electrical circuit breaker panels. The circuit breaker panel feeds into the relay panel, with the relays within the panel acting as a switching device for the circuit. Many relay panels can be fed by both 120V and 277V circuit breaker panels, and relay groups can be fed by different voltages within the same panel. Each relay can be individually programmed through the system controller or the relay panel.

The relay panels provide line voltage control of the lighting loads. Relay panels allow for a single circuit to feed into several relays, and allow multiple circuit breaker panels to feed into a single relay panel. While relay panels can be programmed or controlled by a system controller, they can also operate without the system controller. The relay panels typically have status indicators for the relay outputs, dry contact inputs for program override purposes, and inputs for monitoring devices, such as photocells and occupancy sensors.

In a multi-story facility, there may be a relay panel on each floor controlling all the lights on the floor. Each room on the floor has a local switch and there is also a master switch for the whole floor. The master switch for the floor may be programmed to turn lights on at 7AM and off at 6PM; between 6PM and 7AM, the system may repeatedly perform an *Off* sweep to turn lights out where the programming has been overridden.

6.2.2 Occupancy Sensors

Occupancy or motion sensors are devices that sense the presence or absence of people within their monitoring range. Unlike scheduling controls, occupancy sensors do not operate on a time schedule, they merely detect whether a space is occupied or not. They may be used in restrooms, utility rooms, conference rooms, coffee rooms, locker rooms, and many other spaces. Typically, the sensor and a control unit can be enclosed in one unit, such as a wall box, but for larger facilities, the sensor is tied to a relay panel. The control unit or the relay is programmed to turn lights on when the presence of people is sensed by the motion detector, and may be programmed to turn the lights off when the space is unoccupied for a predetermined time period. The sensitivity of the sensor may also be adjustable.

Figure 6.2 The emergence of LEDs.

There are several types of motion sensors available, including passive infrared (PIR), active ultrasound, and hybrid technologies, such as combinations of PIR and active ultrasound, or PIR and audible sound. These sensors are typically used in locations such as hallways, lobbies, private offices, conference rooms, restrooms, and storage areas.

Ultrasonic sensors emit high frequency sound waves (ultrasonic) and sense the frequency of the reflected waves as they return to the device. Movement in the area where the waves are emitted changes the frequency of the reflected waves, causing the sensor to turn the lights on. Ultrasonic motion detectors provide continuous coverage of an area and are best suited for use in open areas, such as offices, classrooms, and large conference rooms. Mechanical devices that produce vibrations or changes in airflow, such as HVAC systems, can trigger ultrasonic occupancy sensors and cause lights to turn on.

PIR sensors detect radiation, that is, the heat energy that is released by bodies. They are labeled *passive* because they only accept infrared radiation and do not emit anything. PIRs operate in a line of sight and have to see an area, so they cannot be obstructed by open area partitions or tall furniture. PIRs use a lens to focus heat energy so that it may be detected. However, the lens views the covered area through multiple beams or cones and may create coverage gaps. Any objects that prevent the sensors from seeing portions of its designated area will cause the sensor to assume the area is unoccupied and turn the lights off, even when it isn't.

Careful placement of occupancy sensors is required to prevent false (nonintruder caused) alarms. Occupancy sensors should not be mounted in the direction of a window. Although the wavelength of infrared radiation to which the chips are sensitive does not penetrate glass very well, a strong infrared source such as from a vehicle headlight or sunlight reflecting from a vehicle window can overload the chip with enough infrared energy to fool the electronics and cause a false alarm. A person moving on the other side

Figure 6.3 Using lighting for safety and egress.

of the glass however would not be seen by the device. Devices should not be placed in such a position that an HVAC vent would blow hot or cold air onto the surface of device. Although air has very low emissivity (emits very small amounts of infrared energy), the air blowing on the plastic window cover could change the plastic's temperature enough to fool the electronics.

Other technologies and approaches to motion detection include sensing audible noise. Hybrid sensors (PIR and ultrasound, PIR and audible) offer the most effective occupancy detection and have maximum sensitivity without triggering false detections.

The placement of occupancy sensors is key to their working properly. Sensors can be mounted on walls or ceilings, and using multiple sensors can sometimes provide more accurate detection, especially in large or irregularly shaped areas. Occupancy sensors must be able to detect motion in their assigned space while ignoring mechanical vibrations and other false signals. Any malfunctioning of an occupancy sensor could be dangerous, especially if the area were a stairwell or other location where illumination is important for safety. Occupancy controls are best used in applications where occupancy does not follow a set schedule and is not predictable.

6.2.3 Dimmers

Dimmer modules manage low voltage switch and line voltage output controls of the dimmer's lighting loads. Stand-alone dimmers typically have status indicators, analog inputs for photocell or occupancy sensors, diagnostics, and are able to optimize responses for different types of lighting fixtures. Dimmers can be used for specific spaces, such as areas with audio-visual presentations, or throughout the total system for managing large facilities.

Like occupancy sensors, dimmer switches connected to a relay panel. Preset dimming controls from a relay panel provide predetermined dimming for several channels or loads. Presets are tamper-proof, that is to say, they will not allow anyone except the approved and authorized lighting control personnel to override the presets.

Dimming can be used to implement several energy savings strategies. For example, lights can be dimmed when the demand for electricity exceeds a predetermined level, possibly as part of an overall load shedding policy. Such reductions are typically unnoticeable for most users. Another example involves fluorescent lamps. The output of fluorescent lamps decreases over the life of the lamp (the expected depreciated output may be used as an initial design factor). Dimming can be used with new fluorescent lamps to produce the desired light level and then gradually manage the lighting level over the life of the lamps, to produce both a constant level of light output as well as a longer lamp life.

6.2.4   Daylight Harvesting

Photoelectric controls are designed to strategically use daylight to reduce the need for artificial lighting, a process called daylight harvesting. They may be located in perimeter offices, atriums, hallways, or in areas with sky- lights. Ambient light sensors measure natural and ambient light, and based on the amount of natural light, adjust the lighting to maintain a constant light level. In some spaces, manual or automatic blinds, or other means of reducing the direct solar exposure glare, excessive light levels, and heat gain, can be used to supplement the photoelectric control. These may in- clude motorized window shades or blackout shutters.

Proper daylight harvesting design not only includes providing enough daylight to an area, but doing so without any undesirable side effects, such as heat gain and glare. Successful daylight harvesting designs will incorpo- rate shading devices to reduce glare and excess contrast in an area. Window size and spacing, glass type, and the reflectance of interior finishes must be taken into account as well. Despite all of these design considerations, day- light harvesting provides little benefit without an integrated electric lighting system, because of the increased thermal loads from the sun. The electric lighting and thermal loads must be reduced while simultaneously increas- ing daylight to an area.

6.2.5   Ballasts

By definition, an electrical ballast is a device that limits the amount of cur- rent in an electric circuit. In electrical gas discharge lights, such as fluores- cent and neon lights, ballasts control the current flowing through the light.

Incandescent light bulbs produce light by running electricity through a metal filament inside the bulb, which heats it and causes it to glow and emit visible light. When fluorescent lights are turned on, electricity flows to two electrodes on opposite ends of the lamp, causing them to heat up. The electrodes, which are very similar to a filament in an incandescent bulb, then become hot and emit electrons which collide with and ionize noble gas atoms inside the bulb. This creates a voltage difference between the two electrodes, causing electricity to flow between the two electrodes, through the gas in the tube. These gas atoms become hot which vaporizes the liquid mercury inside the tube.

The mercury vapor then becomes excited and emits ultraviolet light, which hits a white phosphor coating that converts the ultraviolet light into visible light. Due to an effect known as *avalanche ionization*, without a device to regulate the electricity flowing through the bulb the gas would continue becoming more excited and higher intensity light would be emitted until the light failed. Thus it is necessary to have a ballast to regulate current through the gas. Modern ballasts supply the electricity needed to start the lamp and produce light, and then regulate the current so the lamp will produce the desired light intensity.

There are two main types of ballasts: magnetic and electronic. Magnetic ballasts use electromagnetic induction to create the voltages used to start and operate fluorescent lights. They contain copper coils which produce electromagnetic fields that control voltage. Magnetic ballasts, which have been used in fluorescent lights since their origin, are considered outdated and are being phased out by newer electronic ballasts. Electronic ballasts use solid state circuitry to control voltage to the lamp rather than magnetic coils, which makes them more energy efficient. One of the main problems with magnetic ballasts is that, while they can control the current to the light, they cannot alter the frequency of the input power. Because of this, the lamp illuminates on each half-cycle of the input power, causing the lamp to flicker and produce a low humming noise. This flicker can cause eye strain and headaches in some people, and the humming can be bothersome and distracting. Electronic ballasts can control the input frequency, eliminating these problems. Another advantage of electronic ballasts over magnetic ballasts is that one electronic ballast can control more than one lamp, allowing for multi-lamp fixtures to be controlled by a single ballast.

From these two main categories, fluorescent ballasts come in three different types: rapid start, programmed start, and instant start. Rapid start ballasts start lamps by simultaneously providing voltage to the electrodes and across the lamp itself. As the electrodes become hotter and emit more electrons, less voltage is required for the lamp, and eventually the cathodes will become hot enough to ignite the lamp on their own.

Programmed start ballasts are more advanced versions of rapid start ballasts. They have preprogrammed start-up sequences designed to give superior longevity to their lamps. Rather than supplying simultaneous voltage, programmed start ballasts first apply voltage to the electrodes to heat them up for a short interval, then apply voltage to the lamps. This helps to avoid a common problem in fluorescent lights called *tube blackening*, which occurs when the electrodes are damaged from voltages without sufficient heating. Programmed start ballasts have the longest lamp lives and are best used in locations with lights that are constantly being turned on and off, such as restrooms.

Instant start ballasts start lamps by providing a high voltage directly to the lamps without preheating the electrodes at all. Because there is no heating time, light is produced within 50 milliseconds, thus giving it the name instant start. Instant start ballasts have the highest energy efficiency of any ballast and the lowest cost, but they suffer from emissive material defects like the rapid start ballasts. Instant start ballasts are best used in lights that are not turned on and off very often.

For certain applications where a light is intended to be constantly on and off, such as a flashing light, there are ballasts which will keep the electrodes heated even when the light is off, which greatly increases the life span of the light.

6.3   Integration into Building Automation Systems

Lighting systems provide a life safety function, assisting in security or lighting evacuation pathways from a building. Lighting systems may be integrated with fire alarm systems, security systems, or emergency power generators. In the case of a fire alarm or loss of normal power, the lighting control system may turn on key emergency lighting fixtures.

Data and information from a lighting system is also an integral part of an overall energy strategy and at some facility or business level needs to be considered with HVAC systems, metering, and building plug loads. Monitoring the number of hours that the lights are operated and the number of times lights are turned on provides information in estimating lamp life, which can be used to schedule group relamping.

Lighting systems have a communication protocol called *digital addressable lighting interface* or DALI for ballasts and relay switches, which was developed in the 1990s. For a DALI implementation, each lamp uses dimming ballast and each lamp has its own network address. DALI uses a two-wire low-voltage wiring scheme in a bus topology, with the bus providing both power and control signals to the ballast.

DALI is an open-source protocol based on standard IEC60929, which specifies performance requirements for electronic ballasts. Each DALI controller (called a busmaster) can handle up to 64 addresses and 16 groupings. Because the DALI protocol is solely for use by lighting control systems its integration to other systems may require a protocol translation with systems using BACnet or LonWorks. DALI has been extended into shading control, and wired network connectivity has been supplemented with wireless connectivity.

Programming lighting control systems use smart technology infrastructure at the higher levels of the control system. Technology evolution is introducing increased use ofstructure cable, Ethernet, and TCP/IP protocols.

6.4   Emerging Lighting Systems

The relentless penetration of IT has a long history of changing building systems; transforming analog phones to digital phones, analog surveillance cameras to IP cameras, IP-enabled access control, IPTV as well as a host of other systems. The next building system to transform to an IT structure is low voltage LED systems. The impetus for an IT structure is linked to the fact that LEDs are low voltage light sources. One way of providing low voltage is installing AC power and converting it to DC; this will work but adds costs, additional points of failure, and generates additional unwanted heat.

Meanwhile, the IT industry has been providing low voltage DC power via Power over Ethernet (PoE) for over a decade. In 2003 an IEEE standard was published allowing low voltage power (48 VDC) to be transmitted over an Ethernet Category 5 Twisted Pair cable. With the initial standard, the maximum that can be delivered to a powered Ethernet device is 15.4 watts, which is sufficient to power many low powered devices. Since its inception, the wattage of PoE has increased, and is now able to provide as much as 100 Watts for some devices.

PoE has several benefits:

▶ It costs less: It is much less expensive to provide a PoE network port than to install conduit, wire, a backbox for an AC outlet, a transformer for conversion, and pay for an electrician. PoE significantly reduces the cost of installation and construction. As an example, Purdue University installed over 1,100 PoE wireless access points across campus, and saved $350 to $1,000 per location by not having to install typical AC power. Others have estimated that the electrical cost to provide power to a device is about $864, while the cost of a PoE network port is around $47-$175.

▶ PoE increases reliability: PoE centralizes power distribution. Instead of a power outlet at each local device power is now distributed from the telecom rooms. Centralized power makes it easier to provide uninterruptible and emergency power for critical hardware such as LED emergency lighting, thus increasing system reliability and uptime.

▶ End devices can be monitored and managed: Network switches provide management tools such as the Simple Network Management Protocol (SNMP), which allows staff to manage the end devices, including power to the end device. You can remotely turn the device on or off, change lighting levels, colors, schedules, and monitor energy consumption.

▶ Moves, addition, and changes are easier: PoE allows for asier building renovations and rearranging of spaces since devices only need one ca-

ble. It's easier to install devices on walls or ceilings and to setup temporary installations.

▶ It's an international standard: PoE is marketed and deployed worldwide, allowing manufacturers to avoid supplying different power cords
for different countries, and eliminating the need for installers to worry
about power cords. Manufacturers, contractors, building owners and
designers can deploy a uniform solution around the world.

▶ Less high voltage is used in the building: PoE means that more low voltage distribution is used to power devices, and less high voltage is used
throughout a building. This results in a safer environment and lower
power consumption.

IT-based LED lighting systems can allow for additional sensors, such as
occupancy, temperature, automated daylight harvesting based on ambient
light levels, and passive motion sensors, with the data points shared with
other building systems where appropriate. With the data from the lighting system and the building BMS, there is opportunity to create a rules-
based lighting system, similar to HVAC's fault detection and diagnostic rules,
where real time lighting and environmental data can be used to optimize
the lighting system or can be shared with other building systems.

The use of DC is important. In most buildings, including our homes, we
are surrounded by devices and equipment that internally operates on direct
current (DC). We plug these devices into a typical alternating current (AC)
outlet, but the AC is then converted to DC, and each conversion creates an
energy loss. In addition, many newly constructed buildings are deploying
renewable energy sources such as solar or wind which can generate DC
power. With the large number of DC powered devices in buildings, and
with DC generation now utilized in many newer structures, the addition of
a DC LED lighting system adds to the consideration of distributed DC power
in buildings and the maximization of the use of DC power generated by
renewables.

Researchers at Carnegie Mellon University have published a paper
evaluating the cost of energy for lighting systems and concluded that a DC
grid is far less expensive powering LED lighting. Researchers say DC power
could save $24,000 a year in a 48,000-ft2 building lit by solid-state lighting
(SSL).

While the approaches to IT-based LED lighting systems are sound, they
do come with a few issues:

▶ IT contractors can certainly install an IT system, but they are neither
electrical nor lighting contractors. They will need training on installing

lighting fixtures and handling stringer supports, mounting downlights in ceilings, or installing ceilings supports for pendant fixtures.

▶ Can IT understand required light levels, light distribution, contrast, color rending, luminous flux, luminous intensity and the lighting needs of particular spaces? IT contractors may need to partner with a lighting company or develop in-house expertise. For a typical lighting control company, the reverse may be true, that is the lighting companies will need to team with an IT contractor or develop internal IT resources.

▶ Also, many times a lighting system for new construction may go beyond a lighting system to include a comprehensive solution incorporating motorized shading and sensors for sun tracking and thermal loads. These systems have to be integrated to optimize the thermal load from the windows with shades so as not to start the HVAC system cooling, as well as optimize occupant light levels. That type of complexity is beyond simply providing low voltage DC to an LED.

▶ Light Emitting Diodes (LED) are semiconductors. They can be configured in an analog mode to be used primarily in older lighting systems, but new systems are likely to be all digital. Digital means an infinite number of colors, scenes, brightness settings, and control. IT-based low voltage LED lighting systems provide energy efficiency, long lasting LED lights, the use of DC power, the opportunity for additional sensors and data points to assist in managing the building's performance, and leveraging the IT network to deploy the lighting systems. The low voltage LED lighting system seems to best reflect the goals of the building industry regarding building control systems.

We typically don't think of the building's envelope as having much need for automation. However the fenestrations—specifically the windows— have a critical role in relation to energy related thermal loads and lighting in building spaces. The sun provides heat and light, and affects the thermal loads and lighting levels in building spaces. The comfort and productivity of the occupants is very important including thermal comfort and visual comfort such as brightness, glare, and shadows. Attention to windows has resulted in automation of windows and window shading devices.

6.4.1 Interior Shading

Until a few years ago most experience with interior shading in a typical commercial office space dealt primarily with the impact on audio-visual presentations in meeting rooms. Often, blackout shades eliminated light and glare, and attendees experienced almost a movie theater type of lighting environment. Interior shading today is focused on energy conservation, specifically addressing the reduction of heat gain from the sun and daylight harvesting.

Interior shading tends to be less effective than exterior shading and window treatments for energy conservation, because interior shading deals with the sun's heat after it has already entered the building. Some of the effectiveness of interior shading depends on the color of the shade; light colors reflecting some of the sun's heat, dark colors absorbing the sun's heat. Except for some energy-related benefits interior shading functions best when it is filtering rather than blocking light and controlling glare.

Manual interior shading further decreases the energy effectiveness of the shading because it depends on a user's behavior, which may be intermittent. Automated interior shading uses motors to move the shade in place, with the control of the motor determining the overall automated functionality of the shade. Automated shades can respond to sun sensors, switches, a schedule, or a specific lighting condition.

When shading is deployed building-wide the network topology of the shading devices is analogous to a BAS system; that is individual shade motors are connected to a controller, which in turn are connected to a BACnet network. Eventually it connects to an IP network which includes a server connected for the management and monitoring of the shades. The shade server will typical have read and write capabilities for all points and values, and provide data on shade position and overall system operation.

6.4.2   Exterior Shading

Exterior shading comes in many flavors. Shading can be attached to the building skin or designed into the building; for example. windows can be set back deeply into a wall section to provide some shading. Both methods need to address exterior aesthetic concerns. Overall, exterior shading may be more aesthetically pleasing than interior shading which tends to have shades in a variety of positions. Exterior shades can be fixed or adjustable. Automated adjustable shades provide the most effective means for energy savings and occupant comfort.

Automated shades, either interior or exterior, have higher initial and ongoing maintenance costs than nonautomated shades, but, are much more energy efficient. So the financial calculation is really a comparison of additional lifecycle costs of automation versus increased lifecycle energy savings.

6.4.3 Electrically Switchable Glass

Electrically switchable glass goes by many names: smart glass, smart glazing, or smart windows. It is basically glass or coatings that change light transmission properties when voltage is applied. There are a variety of technical means to accomplish this including electrochromic, suspended particles, and liquid crystals.

When voltage is applied to electrically switchable glass the devices or coatings change to tint and absorb light. Depending on the underlying technical means, either a one-time or constant electrical current is needed to activate. The coatings or devices return to clear when current is interrupted or polarity of the voltage is reserved.

The clear to tint or tint to clear change can occur in just seconds or a few minutes depending on the technology. The tint level can be controlled manually or automated via integration into a BAS system. The control options vary with manufacturers; some being able to go from clear to tint back and forth, with others having some intermediate levels of tinting. Much like the motorized shades, the electrically switchable glass can be manually operated via a switch or automated based on light sensors, schedule, occupancy sensors, lighting control or thermostats.

Electrically switchable glass is not new and you may have used or seen it. It's been used in interactive displays in museums, outdoor displays, privacy glass, projection screens, and in windows on planes, trains, and automobiles. The rearview mirror in your car may be using one of the underlying technologies in electrically switchable glass.

The issues with electrically switchable glass include installation cost, the limitations on the type of windows offered by some manufacturers (i.e., not applicable to operative windows), the degree of transparency of the glass, switching speeds, and the ability to control intermediate light transmission states. Prices on some of the electrically switchable windows are coming down as companies' ramp up their manufacturing to meet what they see as a huge potential marketplace segment in energy conservation.

6.4.4 Automation Issues

The automation issues with shading may seem simple and straight forward, but, they are not. There are multiple effects we try to optimize with shading and they are interrelated. For example, daylight harvesting may allow us to dim lights but also affects heat gain, possibly occupant comfort, and productivity due to increased glare or brightness. Shading done properly reduces the demand for cooling and provides a modification of the lighting to a space that improves the amount and dispersal of the lighting.

Several manufacturers of motorized shades have addressed this through shading management software, which optimizes the position of the shade based on multiple criteria. These include sun position, solar intensity, BTU Load, readings from indoor and outdoor photo sensors, and radiometers. Some software is able to calculate the optimum shade position every 60 seconds and position the shade accordingly.

There are three systems that need to act in tandem: lighting control systems, the HVAC system, and the shading system. The lighting and HVAC control systems involve energy consumption, so optimal operation should take into account the cost of energy for both systems (i.e., saving money on dimming lights versus the additional cooling due to increased heat gain). Complexity increases when you start to consider systems schedules, sun sensors, occupancy sensors, room temperature, and time of day. Each deployment of automated shading will have different variables to consider, so one generic solution is unattainable. The questions to be raised are what will automatically trigger shading? Will shading be used with daylight harvesting? What level of integration do we need between the systems? And finally, if we are to integrate the systems, what is the sequence of operation among the three systems?

CHAPTER 7

Contents

Data Analytics

Astute facility managers try to extract data from building systems and transform that data into actionable information in order to better manage their building. To most, actionable information means trending the data to observe patterns and exceptions or producing enterprise reports on energy or alarms. What has emerged over the last several years, however, are software analytic applications for building systems, allowing for deeper and more detailed analysis. The basis for many of these software applications is fault detection and diagnostics (FDD), a methodology developed by national research laboratories and relevant government agencies.

FDD has taken actionable information and system intelligence to a new level. There are now a number of credible companies in the marketplace that offer FDD-based tools with case studies of implementations that resulted in optimal system performance, energy savings, and improved operations. For those that understand the return on investment from commissioning building systems and (given the normal degradation of the performance of a HVAC system) FDD-based software tools, FDD becomes a real-time continuous commissioning operational tool. What follows are a description of FDD and some issues in utilizing FDD software tools.

7.1   Overview

While FDD can be used for other building systems, it focuses primarily on HVAC systems. HVAC systems are one of the more complex and energy-consuming systems in a building involving different processes and the interaction of different types of equipment. We measure the performance of an HVAC system in several different ways: indoor air quality, energy consumption, and thermal comfort.

FDD is based on research of faults in HVAC systems and the development of hierarchical relationships and rules between the different equipment and processes that make up the HVAC system. For example, a chilled water plant supplying chilled water to air handling units is a relationship; the chilled water plant is the single *source* and the air handling units are multiple *loads*. Another relationship is an air handling unit delivering supply air to terminal units; the air handling unit is the single source and the terminal units are the multiple loads. In an HVAC system this relationship between source and load can be via air or water. It is these relationships and the rules within the relationships that are at the core of FDD.

These hierarchical relationships are used to collect raw data, apply a set of rules and identify faults. For example, there is a set of rules for systems consisting of a chiller, a boiler, air handling units receiving hot and chilled water from the boiler and chiller, and terminal units receiving supply air from the air handling units. A different set of rules would be applied if there was staged heating and cooling directly at the air handling unit or for single-zone air handling units. There are also different rules for the same equipment based on the state of the equipment; for example a chiller will have a certain set of rules when it is off, another set of rules at start-up and still another during its steady-state.

An example from the US National Institute of Standards and Technology demonstrates how FDD can be applied. In this case it is FDD developed for residential split-system heat pumps. The US Department of Energy implemented a regulation in 2006 requiring a 30% increase in the minimum *seasonal energy efficiency ratio* (SEER) for central air conditioners. Equipment manufacturers made the improvements to their equipment and at the same time NIST's HVAC&R Equipment Performance Group developed a FDD approach for residential heat pumps to assess and assure the improvements in the equipment. NIST identified six common faults in the equipment: improper refrigerant charge, improper indoor airflow, incorrect outdoor airflow, flow restriction in the refrigerant liquid line, noncondensable gases in the refrigerant and reversing valve or compressor valve refrigerant leakage/bypassing. They took a look at fault-free and faulty performance characteristics for both heating and cooling modes. The research resulted in a set of FDD rules that will be used in the commissioning and the detection of sensor failure for residential split-system heat pumps.

In a study on ongoing commissioning (an element of which is system diagnostics) Lawrence Berkeley Laboratories, showed an average energy savings of 10 percent and as much as 35 percent in some cases. Many of the FDD tools go beyond just identifying faults in building systems and can provide guidance on what the root cause of the fault may be. This is information that is valuable to the facility engineer and saves time. Some of these tools can also monetize the fault, where there is some indication of the severity of the problem or its wasted energy. The monetization of the faults tends to rearrange the priorities and urgency of work orders.

7.2   Issues and Concerns in Implementing FDD

Lack of Data: FDD needs data from the BAS systems. If there are not enough sensors, the sensors are inaccurate, or the building has a legacy control system, there can be issues with obtaining the data required.

Rules Specific to Building Systems: The rules apply to specific HVAC relationships and equipment, and building owners need to be assured that their specific building systems are addressed by the FDD software application or can be developed. Many FDD software products start with FDD rules developed by NIST that are then augmented with rules developed by others or by the companies themselves.

How to Handle the FDD Information: Facility Management organizations need to decide how best to handle the FDD information. A fault indicates that the system may be operational, but, is not performing optimally. Faults should not be treated as an alarm. An alarm is a condition meaning immediate response. Faults need to be addressed. Some facility management groups set faults aside, study the remedies, and manually prioritize the faults. Monetizing the faults helps in prioritizing the faults.

Using the Diagnostic Data: Many FDD-based software tools can provide information to the technician or engineer regarding potential corrective actions. This information should be integrated into the work order system, which may be one application in a whole suite of facility management applications, in order to effectively utilize the information.

Prognostics Data: While FDD seems inherently capable of providing prognostic data (which it can analyze for fault conditions or degradation faults, predicting when a component will fail) very little has been developed in the area of predictive maintenance. In addition, prognostic data would allow for more proactive condition-based maintenance which may be a better approach for facility management organizations that are reactive and corrective.

Figure 7.1

Lack of Applications For Emerging Systems: FDD routines do not currently address newer on-site energy sources such as solar, wind or geothermal, or touch on power management or demand response.

SaaS and Hosted Systems: Many of the current offerings are provided as Software as a Service (SaaS). This can be an issue with corporate IT departments because of the need to pierce the IT firewalls and security. However, some facility management organizations see it as an advantage because it means less involvement and dependence on the IT department.

Alternative Ways to Deploy FDD: At some point in the future control manufacturers will integrate FDD routines into their controllers, starting with the large equipment such as chillers.

Keeping a significant energy consuming system such as the HVAC running at optimal performance is challenging. Many times failures or suboptimal performance goes unnoticed for long periods of time. Case studies from companies vending FDD-based software services can show energy savings in the 10% to 35% range with the capability to correctly identify faults and the primary response 95% of the time.

Software based on FDD is a new class of tools for building owners adding some "smarts" to a smart building. It's not difficult to imagine similar tools for other building systems and the potential for enhanced intelligence built into tools for facility management.Recently, the best use of an analytic software application for building systems has been fault detection and diagnostics (FDD) for HVAC systems. There is research including case studies with verified results showing that analytic software reduced energy consumption, improved the efficiency and effectiveness of building operation, and reduced building operation costs. Once used, FDD becomes a core operational tool for many facility management organizations.

Despite the impressive progress with FDD, the industry is in its infancy in deploying data analytic applications in buildings. If analytics for the HVAC system have provided outstanding outcomes, we need to take that template to other building systems. Such applications are based on rules

Figure 7.2

of how the system should optimally operate, generally obtained from the original design documents, and monitoring key data points in near real-time. You essentially compare the real-time data with the rules, and if the data adheres to the rule, the system is fine. If not, the system is not running optimally and has a fault. For those systems that are not process based, applying analytics generally uses statistical monitoring of key performance indicators (KPIs) to monitor outliers. This may not provide diagnosis of an issue, but it can identify faulty equipment for preventative maintenance.

7.3   Guest Industry Experts

To get a glimpse of how we could take analytics to other building systems, and possibly spark ideas from others, we asked some the best known industry experts to provide examples. What follows is their examples:

7.3.1   Lighting Systems

Mike Welch, Control Network Solutions

The key to analytics for lighting systems is addressable networkable light fixtures communicating with an open standard protocol such as DALI® or IEC 62386. This allows the acquisition of data and command points which can develop into value information. This value information for standard lamp fixtures can be as simple as:

▶ Lamp failure

▶ LED driver/Lamp Ballast failure

▶ LED overvoltage

▶ Mains Failure

▶ DALI network failure

The more complex reporting would be:

❱ Run time

❱ Calculated power per fitting

❱ Dirty lamp fitting versus lamp fitting failing

❱ General lamp fixture failure

❱ Lamp change versus energy consumption cost analysis

There are also analytic opportunities with emergency lamp fixtures. These fixtures typically include a lamp with a rechargeable battery pack unit, and need to be tested regularly by law in most jurisdictions. While these tests are usually performed manually, networkable emergency fixtures can be fully automated, from actually initiating the tests to recording results or sending alarms when failures occur. The data that can be acquired includes:

❱ Battery status

❱ Battery charge level

❱ Lamp status

❱ Emergency fixture availability

❱ Runtime

❱ Mains Failure

❱ Network failure

7.3.2   Water System and Conveyance Equipment

Andres Szmulewicz

For water distribution or irrigation, you would be monitoring power usage and gallons per minute (gpm) of the pumps. If you see the kw/gallon drift up from historical data for the same gpm (i.e., not just a performance curve variance), you know that the pump (while operational) is in need of service because it is becoming inefficient.

For conveyance equipment, you would be monitoring the weight load sensor, drive power consumption, and travel distance. If you see the kw/ft.*lb drift up from historical data, you know that the motor (again, while operational) is in need of service because it is becoming inefficient.

7.3.3 Power Management Systems

Brian Turner, President, Controlco

There are several other systems in the building envelope that benefit from analytics and fault detection, especially when they are integrated into the same architecture. Lighting controls and power management are the next logical systems to include fault detection. While it is a fact that these systems already include some level of fault detection, there is a strong argument to be made that including these faults within the context of the building is a better solution.

Lighting control systems often include fault and system alarms for the devices and controllers. They rarely have algorithms defined to measure how well the sequences are performing, and if the overrides are impacted, the overall energy performance of the building. Similar to HVAC, lighting sequences and control strategies are customized for the building and tenant requirements. It is simply not possible for the manufacturer to predetermine what analytics or fault detection algorithms to include in the system before the sequences are commissioned. Once the sequences are understood, rules can be created that will measure the effectiveness of the sequence as compared to occupancy, HVAC schedules, energy goals, safety, and other related and nonrelated data sets.

Power management is another area of fault detection. Again, the system rules cannot be built until the building sequences for HVAC and lighting are implemented, and time has passed to build a baseline of information. Energy managers largely look at the energy consumption in reports, either through dashboards or spreadsheets, to determine if they are meeting their energy goals. Analytics and fault detection can be applied to look for the same anomalies the energy managers look for, but they do it near real time and inform the appropriate energy manager of the situation. If the proper systems are in place, additional rules can be implemented to automatically take corrective action to avoid new demand peaks or energy records.

7.3.4 IT Infrastructure

Jim Lee, Cimetrics

Many organizations have IT infrastructure that is both centralized (data centers) and distributed (telecommunications/server rooms). Although energy efficiency is sometimes considered, especially during construction, the real value propositions that matter to the IT departmentare reliability and risk mitigation (ensuring uptime). The reliability of the infrastructure depends on a number of factors including environmental conditions, power quality, and IT equipment performance. Fault detection and diagnostics (FDD) can be applied to the reliability problem, but it requires the fusion of data from disparate sources such as HVAC, power metering, electrical switch-gear, uninterruptable power supplies, backup generators, servers, routers and switches. Data is collected via SNMP, Modbus, BACnet, and proprietary protocols. Today's servers can provide management data, including internal temperatures and computational workload metrics, which are also useful inputs to an FDD system. For IT, the benefits of FDD are to predict equipment failures before they occur and to provide insight into the cause of failures that do occur. This requires that FDD analytics operate at the system level as well as at the equipment level.

7.3.5 Demand Response and Refrigeration

Brian Thompson and Chuck Sloup, Ezenics

7.3.5.1 Demand Response

Demand Response is reducing or curtailing their electricity use (load) when requested by the utility grid during periods of high power prices or when the reliability of the grid is threatened. Fault detection is different than demand response but can be used to determine if a demand reduction program has met its bid/goals. If the customer finds that they are not meeting bid/goals, analytics will be able to tell why the goals were not met in order to make changes so that the next DR event can be successful. For this to happen, building systems that consume energy need to be monitored for correct operation and coordinated in a central location. For instance, monitoring the operating modes of a lighting control system in a retail establishment will determine if they system is ready for an automated DR event such as shutting off certain lighting tracks in order to reduce load. This automated load reduction cannot be achieved if the lighting circuit HOA switch has been overridden to the hand position. Therefore the analytics will send a notice to the store team to make the correction to make the system ready

7.3.5.2 Refrigeration

Another benefit of analytics is the ability to monitor systems that are an integral part of the client delivering the product or service to their customers. For instance, monitoring the refrigeration system in a grocery/convenience store will determine if the system is working at peak efficiency and effectiveness thereby preventing product loss due to spoilage. The criteria for peak efficiency and effectiveness are typically defined by the refrigeration case manufacturer. A typical requirement is to achieve a minimum number of defrost cycles per day to make sure the evaporator coil can achieve maximum flow, thereby protecting product from frost buildup.

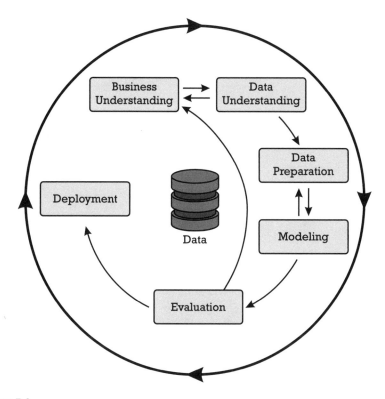

Figure 7.3

Analytic software is a new class of tools for building owners, providing them real-time analysis and diagnostics of their systems and adding some "smarts" to a smart building. As John Petze, Brian Turner, Brian Thompson, Andres Szmulewicz, Mike Welch, and Jim Lee have demonstrated, it's not difficult to imagine analytic tools for other building systems and the potential for enhanced intelligence built into tools for facility management.

Fault detection and diagnostics, like other analytic software tools related to building systems, primarily supports technicians and engineers in the field who are dealing with everyday operational matters and details of building operations as well as the broader issues of complicated systems, advanced technology, and higher expectations for building performance. The fault detection and diagnostics analytic tools provide insights into building systems that help reduce energy consumption, improve building performance and lower costs. Fault detection and diagnostics does just what its name implies: It finds problems within HVAC systems and offers guidance about solving those problems.

7.4   Case Study: Microsoft Redmond Campus

In or around 2010, Microsoft started a project focusing on energy management and improved building operations. The project involved the integration of seven building management systems used on the campus, and the deployment of fault detection and diagnostics across its campus. The campus has 15 million square feet of space in approximately 120 buildings.

The initial pilot of the FDD application consisted of 2.5 million square feet of space, where three potential contractors demonstrated their software, prior to selecting a contractor and deploying the application across the campus. Although Microsoft benefits from the third lowest utility rates in the country, they were able to save over $1 million in energy cost the first year from the use of FDD.

Microsoft invested 10% of the yearly energy costs in the deployment of the FDD application, and had a payback of less than 18 months. Microsoft previously had a process of recommissioning each of its buildings every five years. After one year with the FDD application they are not recommissioning each building every year; saving energy, and operational costs.

Table 7.1

Definition of a fault and problems that can be detected with FDD

A fault is a deviation in the value of at least one characteristic variable from its normal expected behavior. Faults that can be detected through FDD include:
HVAC systems that improperly simultaneously heat and cool
Excessive outdoor air intake and conditioning
Under-utilized free cooling potential
Equipment malfunction (such as broken/leaking valves, broken/stuck dampers, sensors out of calibration)
Systems with the wrong setpoints and operating schedules
Unintentional manual overrides
Lack of energy-saving control sequences (such as chilled water reset)
A bad bearing in a motor or compressor (the bearing then can be replaced before the whole gear box fails and becomes an emergency repair)
Misaligned motor, rotor imbalance, or cracked rotor bar
Dirty filters or strainers
Incorrect refrigerant or oil levels
Pumps with throttled discharges
Short cycling of equipment
Excessive oscillation (hunting) of control points and/or control loop tuning needs
Incorrect fan and pump speeds, pressures, or low flow rates.
Improper building or space pressurizations (negative or positive)
Inefficient boiler combustion
Excessive building peak electrical demand

MICROSOFT Illustrative example of fault detection and diagnosis output (simplified). Microsoft is now collecting 500 Million data point values every day, and using that data to create energy and operational savings. The company expects to reduce energy consumption by 10 percent.

The use of analytic software applications have demonstrated positive measurable results. A study by Lawrence Berkley Labs titled "Automated Continuous Commissioning of Commercial Buildings in 2011, indicated 30% reduction in building total energy consumption and related costs over the baseline; and 30% reduction in building peak demand and CO_2 emissions On top of that are operational savings related to increasing personnel efficiency and effectiveness; facility engineers and technicians being more quickly alerted to a fault in a building system, provided improved information on potential system remedies, and monetizing faults to indicate the wasted energy. Given their track record, analytic applications have been successfully used in lighting systems, electrical distribution, conveyance equipment, data centers, etc.

CHAPTER

8

Contents

Monitoring Conveyance Systems

Elevators, escalators, and moving walkways are key equipment that facilitates the movement of large numbers of people for public transit, airports, theme parks, large buildings, and other venues. Without such equipment, people traffic and the related business operations are significantly slowed.

An example of the critical role that conveyance equipment plays in the operation of these types of facilities is the London Underground, where the famous Tube transports over 1.1 billion passengers per year. The conveyance equipment used by the London Underground includes 426 escalators (one with a 90 foot vertical rise) and 164 lifts, or elevators. This equipment runs 20 hours a day and is able to handle over 13,000 passengers per hour. With that kind of volume and traffic, extremely high reliability of the conveyance equipment is a requirement.

Relying on complaints from the pubic to notify management of an outage in conveyance equipment will not result in high reliability of the systems. Achieving high reliability requires the equipment to be monitored and managed in real time. Such monitoring can provide facility management with the current status of the equipment, as well as the capability to analyze

Figure 8.1

data to discover and predict its performance, use, and the need for maintenance. If a facility has hundreds of conveyance systems, it's going to be impossible to keep 100% of them operational at any given time; however, a smart monitoring system can get you in the high 90% range.

There are a few key indicators of equipment performance that help in managing these conveyance systems. The following are the basic metrics, required data, and applications to keep systems running smoothly.

8.1   Wait Time for Elevators

People hate to wait. If there are long wait times for passengers, there's probably a problem with the elevators related to equipment condition or legacy controls, such as the use of relays. You need to measure wait times during peak and off-peak times and adjust equipment use as needed. Theresa Christy, a mathematician at Otis Elevator says that people get impatient and agitated after 20 seconds of waiting. Studies show that people overestimate

Figure 8.2   Esclator.

how long they've waited in a line by 36%. Remarkably, some building owners that have had complaints of elevator long wait times have determined
that the complaints are the result of boredom, and have responded by making the lobbies more interesting.

8.2   Elevator Speed

Elevator speed is important. One needs to periodically monitor the time it
takes to go from the bottom to the top floor. This is a good metric to measure and observe to see whether the speed is deteriorating, indicating issues
with the equipment.

The speed of an elevator is related to the different type of elevators
(hydraulic, geared-traction, gearless-traction) and how tall the building is.
A three-story building using a hydraulic system elevator has a speed around
100 feet per minute; a geared-traction elevator in a 25 story building may
be 700 feet per minute; real high rise buildings have speeds over 1,000 feet
per minute. During the recent renovation of the Empire State Building,
the speed of the elevators was increased to 20 feet per second to get more
people up to the observation deck. Passengers now rise 80 floors in about
48 seconds (without seatbelts).

8.3   Temperature and Humidity in The Machine Room

Bad affects of temperature and humidity are not related to equipment use,
but, to the conditions experienced by the equipment controller. Many machine rooms have inadequate ventilation; they're too hot, too humid, and
the outside air is not filtered. You don't want the elevator drive and control
system to overheat, fatigue the electronic controls, and possibly increase
breakdowns. Even if you don't have breakdowns you'll shorten the life
of the control equipment. These rooms need to be treated similarly to a

Figure 8.3 Moving walkways.

telecommunication room with adequate cooling and air quality, and sensors for environmental and security aspects being remotely monitored.

8.4 Energy Consumption

On the initial industry push for building energy conservation and sustainability, conveyance equipment was pretty much disregarded by the green certifications. Generally elevator energy consumption is about 5–10% of total energy consumption for a typical 5 to 30 floor building, modest compared to HVAC and lighting systems, but, still significant. In transit organizations and airports where you have hundreds of individual conveyance units, one would expect that percentage to be considerably higher. Organizations need to submeter the power provided to the conveyance equipment. This allows for understanding the energy consumption of the equipment, and its performance. The energy data can be rolled up into an enterprise energy management system and will provide some granularity to overall energy consumption of the organization.

8.5 Use Video Cameras

It may seem odd to use video cameras to capture activity of conveyance equipment but their use has two benefits. One is that large transit organizations or airports may be liable for accidents involving passengers using the conveyance equipment. Video provides a record of what exactly happened to cause the accident and helps in determining any liability. The second benefit is that video can also help in managing the performance of the equipment. If the equipment stops or has a fault, video before and after the event can be used to help diagnose the problem.

8.6 Relevant Conveyance Data

The data that can be acquired will be limited by the configuration of the controller. Controllers for conveyance equipment are Programmable Logic Controllers (PLCs) and are configured using modules for different aspects of the equipment. Of course, alarms would be a top monitoring priority. Table 8.1 shows some basic (nonexhaustive) data that should also be collected.

8.7 Applications

Acquiring relevant data from the conveyance equipment is one thing. Transforming the data into useful information is another. It involves identifying individuals or groups within the organization that would be interested

Table 8.1

Elevator Data	Escalator
Car position	Running Up
Direction	Running Down
Car Load	Sleep Mode
Door Status	Out of Service
Total number of door openings	Emergency Stop
Number of runs per car or call	Chain tension
Up and down hall cards	Motor speed
Out of Service	Oil Level Drive Moter
	Chain Roller Wear
	Electric Noise Generation
	Step Roller Wear

in the information, what their key performance indicators are, and displaying the information in a way that a user can perceive the information, comprehend its importance, and project what needs to be done based on the information.

Typical monitoring applications include:

▶ Real Time Status of Equipment: This should be an enterprise or summary view of the equipment with the capability to drill down to view a schematic of a particular piece of equipment with all current data points.

▶ Alarm Management: This only works if an alarm management plan is developed for how the alarms will be prioritized and classified. Underlying the rules should be an assessment of the potential consequences of each alarm. For each alarm determine the response of the building technician or operator. Identify the roles and responsibilities of each person involved in responding, as well as the work process from alarm notification to resolution and documentation. Hierarchy rules for escalation should also be included. This application should involve integration into a work order system.

▶ Preventative Maintenance: This involves an application with a Facility Management System. Data on the preventative maintenance of the conveyance equipment would be gathered with work orders automatically processed upon key preventative maintenance dates or events.

▶ Equipment Usage and Traffic Patterns: The system should monitor and trend the usage of the equipment by hour, day, and month, and track this information against the people or passengers using or occupying the facility.

▶ Operating Conditions: The system should monitor the quality of power to the equipment, environmental conditions, and security alarms for the machine rooms, as well as energy consumption.

Once you appreciate the role of conveyance equipment in building operations, the monitoring of the equipment will be critical in improving building performance.

CHAPTER

9

Real Time Location Systems

The idea of locating objects and finding people in buildings is not new. Back in the 1990s Real Time Locating Systems (RTLS) entered the market. There are numerous business reasons why building owners would want to track people, equipment, and materials on their property. They may want to better track and manage the assets, equipment, and personnel within the building, increase the efficiencies and workflow of the operations, prevent theft, or improve the response to life safety situations.

For decades, bar codes have been the technology of choice for ID and tracking. However, they can't be changed, need a line-of-sight to be read, have a short lifespan and offer minimal security. Newer technologies offer greater data capacities, higher security, tags that can be modified, and can be read without line-of-sight or contact. They may even be able to integrate into existing wireless systems.

Many wireless tracking systems are based on RFID (Radio Frequency Identification) and are fairly common. If you have been to a library recently you probably have used this technology to check out a book. Library books have RFID tags. Sensors at the checkout counter or

Figure 9.1

the door sense the RFID tag and the system identifies and tracks the book. Walmart and the U. S. Department of Defense legitimized the technology and furthered its adoption by mandating its use in their supply-chain processes.

Other RTLS technology is based on Wi-Fi. The technology may the basis for the RTLS or the Wi-Fi network can be networked with active RFID systems. The Wi-Fi system uses wireless access points connected to a wired IT network and adheres to IT standards using Ethernet and radio frequencies standardized by IEEE 802.11. Initially the density of Wi-Fi access points in buildings had to be significantly increased to allow the system to *triangulate* the object or person and the Wi-Fi tags. The objects were fairly bulky, though there have been improvements to system accuracy and tag weight.

The technology for a RTLS deployment is dependent on how it's utilized. For large manufacturing or production the use of Wi-Fi may be more suitable given the reach, but, if you're tracking a volume of smaller consumable items, passive RFID Tracking may be more appropriate. What follows is a description of RFID tracking systems and how they can be used in a healthcare environment.

9.1 Tags

9.1.1 Barcodes

Barcode systems use lasers or cameras to read barcode labels. These systems have been around for many years and are less expensive, but, require line-of-sight access to barcode labels in order to be read. They can be time consuming as each label must be found and read. Barcodes are cheap and work well, but they require a human being to perform a scan using line-of-sight.

Figure 9.2

The handheld scanners used to read barcodes work well. One of the advantages of barcodes is you scan only one asset. RFID can scan many assets at one time.

9.1.2 RFID

RFID (radio frequency identification) uses radio frequency waves to sense tags within its read field. An RFID reader emits a signal, which can detect RFID tags that contain data. There are many types of RFID tags and readers, which create read fields of widely varying sizes. This can be just a few centimeters (nfc), to hundreds of feet (active RFID).

RFID is faster and easier for collecting a lot of data in an automated way. Prior to deploying an RFID system, it's suggested that a building owner check the radio frequency in the current environment for conflicts in common frequency ranges. RFID tags are simply radio transponders. They are a small integrated circuit or computer chip which has a very small radio antenna built in. In passive systems, the tag does not have its own power source—it absorbs energy from the system reader antenna that senses the tag. The tag has been programmed with its own unique identification. When the tag is excited and absorbs the radio waves of the reader antenna, the tag sends out its unique identification, which is picked up by the reader antenna. Passive tags are typically smaller, less expensive and have a shorter range. They may be used with low cost assets in retail or manufacturing environments.

In active systems, tags have their own power source (battery) and don't need to use the reader's antenna radio waves to power up and transmit their identity. Active tags have a greater range, can store larger amounts of data and are larger than passive tags.

Tags come in a variety of sizes and shapes to address a variety of uses. The tag can be paper thin to fit inside a book. They also can be directly mounted onto equipment with an adhesive. Tags can be embedded in wrist straps, attached by a pin to clothing, worn on a belt, and even made tamperproof.

9.1.3 QR (Quick Response Code)

QR codes are two dimensional barcodes. Scanning can mean sending a text message, receiving a hyperlink, dialing a telephone number, or sending a message to an email address. QR has two major benefits. First, QR can handle a lot of data (7,089 numeric characters, or 4,296 alphanumeric characters), and second it can be scanned with a smartphone.

9.1.4 Readers and Antennas

Readers have an antenna attached to them. Essentially the reader interfaces or sits between the wireless portion of the system (the antenna) and the headend or system management and administration server. The antenna attached to the reader sends radio signals out that activate tags. It listens for tags to send a response and once it does, reads the data transmitted by the tag and sends it to the reader. Readers can have multiple antennas attached to it. The reader can decipher the signal and send the data to the host server.

9.2 RTLS Host

The host server can have several functions. It can configure, manage, and monitor the readers. It can also host specific application software, such as card catalogue software for a library's RFID system. The host system can act as a middleman between the RFID system and other software applications that a client may have, such as asset inventory, purchasing, patient records, or pharmaceutical supplies.

9.2.1 RTLS Healthcare Example

A hospital needs a fast and effective way of finding people (e.g., patients, doctors, nurses, orderly, translators, or duty staff), as well as locating key medical. The capability to identify persons and equipment in the hospital through the use of RFID addresses this need and provides several important benefits.

For hospital staff, an RFID chip is embedded into a smart card. Applications for the smart card for hospital staff include identifying their location, access control, point of sale applications, parking garage access, and event management. The system would also have the capability to list staff members on duty.

The hospital can tag mobile medical equipment. The RFID system can quickly locate needed medical equipment saving significant time for staff while interfacing into an asset management system.

Patients are provided with wristbands with embedded RFID chips, allowing patients to be automatically identified and located. The tags are also critical as a security measure, providing an alarm if a patient or infant were to leave the hospital premises unauthorized.

These systems can automatically track and time stamp the progress of people or assets through a process, such as following a patient's emergency room wait time, time spent in the operating room, and total time until discharge. Such a system can be utilized for quality control and improves delivery of services.

9.2.2   Administrating an RTLS

RTLS is a technology system used as a tool for organizations, primarily for asset management. There are several business benefits with a current, complete, and accurate facility equipment asset management system:

▶ By tracking the useful life of assets the building owner can understand the return on assets.

▶ An asset management system can automatically provide the history of the asset including all new work orders on the item in order to assist the technician or engineer in repairs and maintenance.

▶ Building owners can analyze the repair and maintenance of similar equipment types or models from different manufacturers to identify optimal equipment performance or common problems.

While an asset management system will be utilized by facility staff, there are other groups or departments that will need to be integrated as well. Purchasing would be involved in receiving equipment and dealing with invoices and payments, as well as coordinating the asset management data such as date of acquisition. The accounting department also has an interest in the equipment assets. Assets may be depreciated or taxed and accounting will want to know where the asset is located.

The naming conventions used for the assets (locations, asset categories) need to be consistent across multiple departments; without such continuity, one could expect increased labor, time, and errors. By using one platform as the database for all facility asset data there is opportunity to integrate asset data in other databases, such as business operations, human resources, and enterprise management performance.

The Data Elements Required in the Asset Database should include:

▶ Country

- Asset Class

- Equipment Type

- Building Location

- Equipment Location

- Make/Manufacturer

- Model

- Serial Number

- Installation Date

An example would be Table 9.1.

The major asset classes would be based on asset management groups. The naming conventions should have a structure where the facility technicians should be able to deconstruct the asset labeling convention for meaning.

9.2.2.1 Location Tagging

Asset management should also track locations. Just like assets, one would assign a unique location code to each space, building, or a specific part of the building. The location tags would have to be uniquely different from the asset tags so that scanning tools can recognize the difference.

9.2.2.2 Applying Asset Tags

If the organization receives an asset from a contractor, the asset should be scanned only for the serial number. You affix asset tags to the equipment when staging the asset for deployment. This requires that asset tags be placed on every asset as soon as it is in the possession of facility staff. Building owners should minimize the number of people responsible for tagging.

9.3   RTLS and Indoor Positioning Systems

RTLS and indoor positioning systems have many similarities. The differences in the systems involve the technology used and the purpose of the system. Most indoor positioning systems use Bluetooth beaconing. Some systems utilize Wi-Fi, but, Wi-Fi is less accurate. The tag used in indoor positioning is a person's smartphone. A ubiquitous type tag may be less secure, depending on the mode of communication between the tag and the Bluetooth beacon.

Table 9.1

Country	Assest Class	Equipment Type	Store Location	Equip. Location	Manufacturer	Model	Serial Number	Installation Date
USA	Mech	RTU	12345	NE corner	Trane	IntelliPak	M123456	9/12/2012
CAN	Electric	Generator	98765				E7891011	10/1/2013
		Lighting		Aisle 15				
	Refrig	Coolers						

The purpose of indoor positioning is to locate an individual or count people within a space. You can also use indoor positioning for business purposes. For example, an indoor positioning system in a shopping mall may communicate to mall shoppers via the individual's smartphone with a coupon from one of the retail stores.

Studies have found that people spend 80–90% of their time indoors, so it's no surprise that one of the most promising technologies for buildings is Indoor Positioning Systems (IPS). An IPS is just a data acquisition system, obtaining information as to location of people or objects within the building, providing data to occupants to assist in wayfinding. It has more potential uses, such as providing valuable information to businesses and building owners, security and life safety functions, and yet in some cases present some legitimate concerns.

Many people are familiar with Global Positioning Systems (GPS). GPS depends on satellites; its use indoors is affected by roofs, walls, and objects in the building resulting in the attenuating, weakening, reflecting, or scattering the radio waves from satellites. While typical GPS may not be suitable for indoor, use GPS and indoor positioning systems can be integrated. Also, some GPS have enhanced the sensitivity and power of their receivers (high sensitivity) and are able to receive some, satellite signals within a building.

Indoor Positioning Systems also have some similarities with Distributed Antenna Systems (DAS) that boost cell phone coverage in buildings. The IPS uses cellphone apps to allow users to navigate through a building and can use cellphone capabilities likeBluetooth to locate the cell phone users. Unlike the real time location systems (RTLS) where corporate users (hospitals, warehouses) bought a system, the DAS deployments are usually business deals between the building owner and local cell carrier. Building owners want to satisfy the needs of their tenants, and the cell carriers want to increase usage. It's likely the indoor positioning systems will involve a commercial arrangement similar to DAS. Questions and concerns will arise about who will manage the system, who pays for what, who owns the data, who profits, privacy and security concerns, and who maintains and upgrades the system.

One approach to the IPS architecture is *Bluetooth Beaconing*. Bluetooth was created in 1994 to replace short cables. Today every smartphone is Bluetooth-enabled and we're all familiar with using Bluetooth to connect ear buds, headsets, printers, game consoles, and keyboards. It is the Bluetooth capabilities of smartphones together with the Bluetooth beacons that can provide the location of smartphone users. In 2010 Nokia introduced an indoor positioning system based on Bluetooth Low Energy (BLE) technology (basically the latest Bluetooth technology operating on low power with low latency in communications). The density of the Bluetooth beacons is roughly the same as a typical Wi-Fi deployment, with accuracy around 0.3 meters (1 foot), with no latency.

There appear to be two general approaches for the Bluetooth beaconing: one where the smartphone pulls data from beacons; the other just the opposite, where the beacons pull information from the smartphones. Several experts believe the first approach is a better plan for privacy and data security. The second approach, with beacons detecting the smartphone, eliminates the need for specialized software on the smartphones and is passive for the smartphone.

While Bluetooth Low Energy (BLE) beacons may be the most widely used technology for indoor position systems, there are other existing technologies. These include Wi-Fi, which is less accurate that Bluetooth. Newer technologies are starting to be developed; these include LED lighting that emits modulated light specific to an indoor location, reading the Earth's magnetic field energy, and embedding different sensors in smartphones. Wi-Fi and Bluetooth BLE are the most common technologies for most IPS deployment.

The IPS system will use a variety of methods including triangulation, signal fingerprinting, and direct association:

▶ Triangulation: Most people understand that cell towers and GPS determine locations based on the distance between a person and at least three radio devices such as Wi-Fi access points or BLE beacons, all of which are known locations. The potential issue here is interference; walls, and large objects such as equipment. IPS fingerprinting overcomes the interference problems of triangulation.

▶ Fingerprinting-based system: This method requires capturing radio signals, such as Wi-Fi or Bluetooth beacons, from a smartphone in spaces within a building. Each sampling of radio signals in a specific room

Figure 9.3   Blue beacons.

creates a fingerprint. Once organized, the IPS can be probed, and the system will have the data and analytics to determine the user's location.

▶ Direct Association: This method works based on the addresses or unique identifiers of Wi-Fi access points or Bluetooth beacons. For Bluetooth, an application on a smartphone apprehends this identifier and the IPS can locate the device. Wi-Fi networks are a bit different because there are multiple access points. Each access point has a Basic Service Set Identifier. The Indoor Positioning System uses analytics to determine signal strength between any access point and a smartphone to estimate the smartphone's location.

9.3.1 Companies in the IPS Space

Recently twenty-two international companies formed the In-Location Alliance to standardize and commercialize this technology. The alliance includes large multi-national companies such as Nokia, Samsung, Qualcomm, and Sony. The list also contains Broadcom, Dialog Semiconductor, Eptisa, Geomobile, Genasys, Indra, Insiteo, Nomadic Solutions, Nordic Semiconductor, Nordic Technology Group, NowOn, Primax Electronics, RapidBlue Solutions, Seolane Innovation, TamperSeal AB, Team Action Zone and Visioglobe.

Some tech giants, such as Google, are pursuing and developing the technology but are not part of the alliance. Apple has is expected to be a major player in indoor positioning with their iOS devices as well.

Some of the companies involved in IPS are high tech companies motivated to extend their existing outdoor mapping applications to the indoor environment where GPS doesn't work well due to signal attenuation. In addition to extending their mapping applications and advertising revenue, the companies have an opportunity to create a treasure trove of data that will assist businesses, primarily retail, in identifying customers and in product placement.

9.3.2 Where Are Building Owners?

While all of this takes place inside buildings and has the potential to provide valuable data to building owners it is not the traditional building control system manufacturers, building operators or facility managers, that are driving the IPS deployments. Without building owners' involvement something is likely to be lost in this approach, as several beneficial applications of IPS can be used to improve building operations. For example, one of the key metrics for building management is data on how occupants are using building spaces. That may include when and where building occupants enter and exit the building, what building spaces occupants inhabit, what time or day they occupy the spaces, and duration of occupancy. Such data can be used to optimize building operations, correlating the occupant data to building systems, energy consumption, space utilization, and even utilizing the data for renovations of existing buildings or design of new buildings. IPS could also be part of addressing life safety events, for example, locating people for safety or security purposes. With facility management organizations just starting to wade into the world of data analytics, IPS is posed to provide data on a building on a level that hasn't existed in the past and will enrich the analytics of building operations.

IPS technology can be used for almost any size building but appears to be particularly well-suited for large commercial buildings, educational campuses, malls, airports and guided tours of museums. If you think IPS is pie-in-the-sky and way off in the future, think again:

Google launched their Indoor Maps and Indoor Location in late 2011. They already have over 10,000 floor plans for a variety of buildings in North America, Europe and Japan, and claim 5-10m accuracy inside these buildings. Some of the mapped clients include IKEA, MGM Resorts, Mission College, Mall of America, Macy's, Home Depot, and Bloomingdales. (See http://maps.google.com/help/maps/indoormaps/).

▶ Walgreens has a partnership with a startup named aisle411. Any Walgreen shopper using a smartphone can view maps of any of 8,000 Walgreen stores and locate products down to the section of an aisle.

▶ Zonith has an IPS offering also using Bluetooth beaconing, while targeting a different market focusing on security, safety, and situational awareness applications. They have an application call Lone Worker, which focuses on large utilities, production plants and commercial buildings that can locate lone workers and keep track of employees for safety purposes.

Some of the applications involve real time location of personnel such as doctors, supervisors, technicians, or tracking team members, as well as assets on missions in dark or crowded locations. Some companies have also used IPS to identify occupancy and adjust energy management systems based on where people are gathered.

Case Study—Museums

Many museums around the world are now interacting with visitors and patrons by acquiring data on their behavior and providing way-finding applications for smartphones. The technology systems utilized include indoor positioning and eye tracking systems. One example of an IPS and eye tracking trial deployment is at the Solomon R. Guggenheim Museum in New York. They have installed beacon technology to develop metrics on how many people go to a specific exhibition, how long they stay and an eye tracking system to determine what people look at and how long they gaze. Based on the data developed, the museum then adjusts exhibits, spaces, schedules, and donor outreach. In many ways it is an approach similar to many retailers and malls; tracking the behavior of their customers to enhanced the relationship between the patrons and the museum. What's happening at the Guggenheim Museum is similar to other museums in the US and around the world.

9.4   Security and Indoor Positioning Systems

Many times we associate IPS with commercial applications, such as retail stores and malls, where an IPS is deployed to identify past or potential customers and used to entice customers to purchase via texting and discount coupons. But IPS has many other uses such as life safety and security, providing an innovative way to locate and communicate with people inside buildings during emergency situations.

The United States Federal Communications Commission (FCC) is looking at indoor positioning to enhance emergency response as well. The FCC has suggested a current baseline for indoor positioning for use in emergency response. One of the concerns is determining vertical location in a multilevel building. An FCC report concludes: "While the location positioning platforms tested provided a relatively high level of yield, as well as improved accuracy performance, the results clearly indicate additional development is required."

Life safety situations and emergencies obviously require a more immediate and critical response. These include emergencies, fires, tornado and hurricanes, and active shooter situations. For the security industry, IPS is a key tool in pinpointing the real-time location of individuals inside buildings that help first responders during the incident and also provide information to assess the situation after the incident.

In addition to locating individuals, IPS based on smartphones can provide communications during an incident. Communicating with people in a building by first responders and law enforcement assists law enforcement in locating a possible shooter. Bi-directional communications can help people in the building to be guided to the safest exit passageways. (It is worth noting that over 70 percent of all emergency calls come from mobile phones). The indoor positioning system can also be used in conjunction with an Access Control System to trigger an alarm notification if a worker or asset enters or leaves certain predefined zones

Many deployments of indoor positioning systems use multiple technologies, both Wi-Fi and Bluetooth BLE. Such installations strive for more precise and exact locations. Multiple technologies also provide some resiliency and reliability in the system. The use of multiple technologies may allow for compatibility with a larger spectrum of support for user smartphone and tablet devices.

For incidents that are occurring with growing frequency in shopping malls, office buildings, schools, and even government facilities, hybrid IPS technologies combining Wi-Fi and beacons offer the best methodology available to protect individuals, reduce injuries, and prevent loss of life.

Hybrid IPS technologies offer the most design flexibility, better accuracy, and a lower total cost of ownership—all of which are great advantages to organizations that have limited security budgets, yet have a strong need to provide a safer environment to customers, personnel, and visitors. Beyond that, by leveraging technology that may already be in place, such as existing Wi-Fi networks or beacons in retail environments, hybrid IPS technologies are faster to implement and easier to integrate with existing security systems. Finally, it makes sense to capitalize on the pervasiveness of mobile devices. The FCC is looking at indoor positioning to enhance emergency response.

9.4.1 Indoor Maps

Indoor Positioning systems do not work without indoor maps. Facility owners will need to survey their facility and incorporate the maps into the facility's app. Maps could be CAD drawings, PDFs, images, high and low resolution PNG, Shapefiles, SVG, and XML. There's a new industry creating those data. Micello recently announced it had mapped 15,000 indoor venues. Google, in addition to collecting its own indoor mapping data is crowd-sourcing maps from its proprietors. Nokia is collecting indoor data. Mapping indoor buildings is being implementedworldwide.

Some of the buildings in the United States where indoor mapping is being implemented include:

▶ Airports
  ▶ ATL Hartsfield-Jackson Atlanta International Airport
  ▶ CLT Charlotte/Douglas International Airport
  ▶ DIA Denver International Airport
  ▶ IAH Houston George Bush Intercontinental Airport
  ▶ LAS Las Vegas-McCarran International Airport
  ▶ MSP Minneapolis-St. Paul International Airport

- ❯ ORD O'Hare International Airport
- ❯ SAN - San Diego International Airport
- ❯ SEA Seattle-Tacoma International Airport
- ❯ SFO San Francisco International Airport

❯ Casinos

- ❯ Caesars Entertainment
- ❯ MGM Resorts

❯ Convention Centers

- ❯ Las Vegas Convention Center
- ❯ Landmark
- ❯ Alcatraz
- ❯ Mall
- ❯ Mall of America
- ❯ Federal Realty
- ❯ Pyramid
- ❯ Regency Centers
- ❯ Rouse
- ❯ The Irvine Company

❯ Museums

- ❯ The Smithsonian
- ❯ 9/11 Memorial and Museum

❯ Sports Venues

- ❯ American Airlines Center
- ❯ Churchill Downs
- ❯ Heinz Field
- ❯ Madison Square Garden
- ❯ Superdome
- ❯ Wells Fargo Center

You may have seen a Google car with cameras and sensors which is used to support Google Maps. For indoor space Google has a cartographer for indoor mapping. The cartographer uses a process called *simultaneous localization and mapping* (SLAM), a technique typically used for mapping new locations. As a backpacker walks through a building, the floor plan is automatically generated in real time, Google says. The wearer also uses a tablet to add points of interest while walking around the building (for example, room numbers in a hotel or the exhibits in a museum).

So we track and locate people and objects outdoors in real-time via GPS. IPS provides for similar tracking indoors. In the near future it may be that there will be few places on Earth where we can't be tracked and identified. The key to successful IPS deployments are going to be: (a) users having the option to opt in or out of being tracked via a smartphone app, (b) thoroughly secured IPS systems, and (c) the demonstration of benefits and value for smartphone users, business owners, building owners, and workers in emergencies.

Contents

Eye-Tracking

We use our eyes constantly and how we use them is important. The study of human eye movement has been around for about 150 years, initially performed by simple observation. Eye movement is important because what people look at and how long they look at it influences their decision-making and comprehension.

Anything with a visual element can be eye-tracked. Eye-tracking has many applications including those for building design and operation. One example is a company that has several *mock supermarkets* or *shopper labs* in order to track eye movements as people wander the aisles of retail stores to determine what items or displays catch their eye. Retailers have limited shelf space and strive to maximize sales and profits, so placing items at a certain shelf height or in specific aisles is important to catch customer's eyes, as well as exposing customers to the highest volume of product.

Eye-tracking can benefit a building, especially its interior design, signage, way finding, and ergonomics for manual controls and kiosks. With eye movement directly related to decision making, we'll start to see more use of research in the design and operation of buildings, touching on the visual structure of the facility, its layout, lighting, colors, and placement of objects or controls. Some of the results could include improved productivity in commercial buildings, wellness in hospitals and enhanced learning in schools.

Eye-tracking is a component of new building concepts and design. For example, with new construction, alternative building facades or spaces can be displayed and viewed as three dimensional Building Information Models. By tracking eye movements and the gaze of design participants (building owners, occupants, architects, engineers, designers and consultants) the information from eye-tracking can provide data on the shape, size, texture, color of the façade or spaces revealing what people liked or disliked. Such data feedback can influence the façade or space design. Eye-Tracking can also complement construction safety, where it is used to identify worker's perceptions of hazards.

Much of today's eye-tracking technology is focused on commercial applications. You'll see eye-tracking used in advertising, software interfaces, retail window design, web pages, and almost anything associated with marketing and selling. Much of the eye-tracking is done for prototypes or draft products or ads, gathering data on how a consumer interacts with visual stimulus to perfect the ad or web page. The basic data evaluates what people look at, and how long it holds their gaze.

There are also beneficial eye-tracking applications that can control computers, monitor automobile drivers or pilots, and even applications allowing paralyzed people to operate wheelchairs via eye movement.

10.1   Eye Tracking Technology

Studies show that you can't simply observe and directly detect what people are looking at. Without eye tracking technology, many studies simply ask questions and observe the person's behavior. However eye tracking technology can generate reliable data. Typical eye tracking technology uses a camera and an infrared light source. The light source is directed to one or both eyes while the camera follows and records the reflection of the light as well as the eye or visual features. While there are other ways of tracking eye movements, the use of video cameras sensing reflected light from the eye is not invasive and generally inexpensive.

Typically, an eye tracking project will first involve the relationship between the eye positioning of the participants, and the scenes are measured and calibrated. Each person has to be calibrated via a valid eye movement prior to using the eye tracker. The participant must look at a number of calibration spots. Eye-tracking will then generate data. Finally the eye tracking data is analyzed and subsequently presented in meaningful framework. Eye trackers are primarily looking at the pupils of the eyes, and corneal reflection.

Many times the camera is remotely positioned to collect the data, such as mounted on a retail shelf or wall. Other methods include a head-mounted eye tracker apparatus. Regardless of the eye tracking method the data is collected and analyzed by a computer. This data is used to identify the direction of gaze, pupil diameter, eye rotation, the motion of the eye related to the head, and blink frequency. Eye tracking will measure the point of gaze (where one is looking) or the motion of an eye relative to the head, eye positions and eye movement. There are a number other of methods for measuring eye movement.

The most widely used current designs are video-based eye trackers. A camera focuses on one or both eyes and records their movement as the viewer looks at some kind of stimulus. Most modern eye-trackers use the center of the pupil and infrared light to create *corneal reflections* (CR). The vector between the pupil center and the corneal reflections can be used to compute the point of regard on a surface or the gaze direction.

Software interprets the data that is recorded by the various types of eye trackers, and animates or visually represents it, so that the visual behavior of a person can be graphically displayed. Graphical presentation and quantitative measures of the eye movement events and their parameters can be used in the analysis. The main measurements used in eye-tracking projects are *fixations* and *saccades*. A fixation is when the user's gaze is relatively motionless on a specific area. A saccade is a quick movement between fixations to another element. There are other eye movement events that stem from these basic measures, such as *smooth pursuit* which allow the eyes to closely follow a moving object and *blink rate* measurements.

Heat maps are the most common visualizations of eye tracking. Heat maps are static representations, which can identify where users focused their gaze. Heat maps reveal exactly where people look. The results often point to useful insights.

Eye tracking generates data regarding what a person views and how long their gaze is. This is part of big data where data is obtained, mined and

Figure 10.1   Eye tracking equipment.

analyzed. The heat maps show the distribution of attention with a color coded map superimposed on the stimulus with an intensity indicator.

The resulting data can be statistically analyzed and graphically rendered to provide evidence of specific visual patterns. By examining fixations, saccades, pupil dilation, blinks, and a variety of other behaviors researchers can determine a great deal about the effectiveness of a given medium or product.

The most common use of eye tracking is with web sites. Traditionally web sites have developed data on clicking and scrolling patterns. Eye tracking provides data on what is eye catching on the web site. Eye tracking can be used to measure search efficiency, navigation, and usability.

Eye tracking is commonly utilized in a variety of different advertising media. Commercials, print ads, online ads, and sponsored programs are all conducive to analysis with current eye tracking technology. For instance in newspapers, eye tracking studies can be used to find out in what way advertisements should be mixed with the news in order to catch the subject's eyes.

One of the most promising applications in eye tracking research is in the field of automotive design. Research is currently underway to integrate eye tracking cameras into automobiles. The goal is to provide the vehicle with the capacity to assess the visual behavior of the driver in real-time. The USA National Highway Traffic Safety Administration (NHTSA) estimates that drowsiness is the primary causal factor in 100,000 police-reported accidents per year. Another NHTSA study suggests that 80% of collisions occur within three seconds of a distraction. By equipping automobiles with the ability to monitor drowsiness, inattention, and cognitive engagement driving safety could be dramatically enhanced. For example, the Driver Monitoring System, also known as Driver Attention Monitor, is a vehicle safety system first introduced by Toyota in 2006 in its Lexus models providing a warning if the driver takes his or her eye off the road.

Since 2005, eye-tracking has been used in communication systems for disabled persons: allowing the user to speak, send e-mail, browse the Internet and perform other such activities, using only their eyes. Eye control works even when the user has involuntary movement as a result of Cerebral palsy or other disabilities, and for those who have glasses or other physical interference.

10.1.1 Examples of Museums

New York's Solomon R. Guggenheim Museum recently deployed an indoor positioning system and an eye tracking system. The devices will enable the museum to send messages about artworks to visitors via their smartphones while at the same time collect details about the visitors. The museum then filters the data to better understand guests' behavior, such as how often they visit, which shows they flock to, and what art they ignore.

Understanding audience behavior enables museums to target marketing for future exhibits or personalize messages to visitors based on their past viewing history. From an educational standpoint, this data can help museums find the most effective tools for teaching their audiences about the art on the walls.

Contents

Distributed Antenna Systems

Cell phones are an important part of the way we live, work, entertain, communicate, and interact. There are almost as many cell-phone subscriptions (6.8 billion) as there are people on this earth (seven billion). In 2013 there were 96 cell-phone service subscriptions for every 100 people in the world (information from the United Nations' telecommunications agency). A subscription doesn't mean that everyone has a cell phone; penetration rates in wealthy countries exceed 100% because some people have multiple phones and cell subscriptions. But even in developing and poor countries, the penetration rate of cell phones is around 90%. Africa has very few fixed-line telephones with a rate of 1.4 subscriptions per 100 people, but has 63.5 cell subscriptions per 100 people.

Several studies have profiled the use, behavior, and bond of a smartphone user and the device. A study of 2,000 smartphone users indicates that the average cell phone user reaches for their phone at 7:31am in the morning, and checks personal emails and Facebook before they get out of bed. Incredibly, 40% of cell phone users admit to feeling lost without their device. Another study indicates that the average smartphone user checks their phone 150 times per day.

With cell phone penetration around 90–100% and studies showing that people spend 80-90% of their time indoors, obviously cell phone coverage in buildings is critical. 80% of voice calls and 90% of data usage is indoors. Some buildings have a poor quality cell phone signal, with the result being dropped calls and spotty coverage. Very large facilities, tall buildings, facilities with a high density of occupants, and buildings with a steel roof or steel/aluminum siding, or brick buildings, as well as concrete and steel joists may affect cell signals. Low-E glass windows, which typically are used to reduce energy consumption by minimizing the amount of ultraviolet and infrared light that can pass through also reflect radio waves and cell signals, creating reception issues with in-building wireless coverage.

Other weak wireless situations may be related to a lack of capacity of cell coverage where demand cannot be met. These facilities may be large public venues such as stadiums, convention centers, corporate offices, multitenant high-rise buildings, university campuses, hospitals, manufacturing facilities, upscale hotels and high-rise condos, casinos and federal and local government facilities, as well as rail and subway systems.

The major issues with cell coverage in buildings are either weak signals from the closest cell tower or lack of capacity. The remedies are different for these two problems. Weak signals are typically addressed by strengthening the existing signal from the nearest cell site or tower, referred to as a *repeater-based* solution. This entails an outdoor antenna on the building pointed toward the nearest cell tower site. The antenna receives the signal from a cell tower and sends it through a cable to a *bidirectional amplifier* (BDA) that strengthens or boosts the signal, then relays and amplifies the RF signal traffic between the remote base station and the mobile radios.

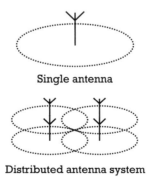

Single antenna

Distributed antenna system

Figure 11.1 Distributed antenna systems.

The issue with increasing cell capacity is typically addressed by installing smaller antennas in the building. These are active systems referred to as a *distributed antenna systems* (DAS). Distributed antenna systems are a network of amplifiers and antennas throughout a facility that are connected with copper or fiber cable to a hub. It provides voice and data wireless service within a geographic area or structure.

The headend of the DAS may allow for multiple wireless providers to connect radios at various radio frequencies. A *neutral host DAS* allows multiple wireless providers to use the network at the same time. The DAS essentially feeds smaller cell antennas that cover the buildings' zones or floors. A distributed antenna system may be deployed indoors or outdoors. For outdoor DAS, some state Public Utility Commissions have require the utility to allow Distributed Antenna Systems in the utility right of way.

11.1 DAS Business Model

Initially wireless carriers deployed Distributed Antenna Systems (DAS) to expand their service into buildings. The carrier funded, installed, and maintained the DAS anticipating increases in their revenues. The carriers realized that their investment in the DAS could be leveraged by leasing the capabilities of the DAS to other carriers, thus creating a neutral host DAS.

The deployment of a DAS can benefit the building owner and the wireless carriers. With the building having a DAS, the tenants and occupants of the building have enhanced cell coverage resulting in reduced complaints and service calls. The DAS also improves the marketability of the property for the building owner by increasing connectivity and increases building safety by improving emergency services. The carriers' benefits include

Figure 11.2 Cell tower.

Figure 11.3 Multiple antennas in a large building.

increased data and voice usage of their wireless networks, leading to additional revenues.

One projection is that DAS installations could grow by 300% by 2017 (iGR Research). Driving this growth is user demand for ubiquitous reliable connectivity, rising bandwidth requirements, and quality of service expectations. The cost for typical DAS systems now range from $5 to $10 million. Worldwide distributed antenna system (DAS) revenue reached $2 billion in 2013. The major DAS leaders are Commscope, Corning Mobile, and TE Connectivity. (Data from Infonetics).

DAS business models vary. In one model the building owner owns the DAS and takes on the responsibility for service, maintenance, and upgrades. Another option is a third party that could provide a neutral-host multi-carrier network; the provider would own, operate, manage, and monitor the network relieving the building owner from having to invest in the DAS. A third option is for the building owner and the third party to split the capital of the DAS with the third party owning the network and being responsible for all maintenance service and carrier relationships.

Generally wireless carriers are hesitant to finance the installation of DAS in office buildings and malls, they look instead at larger venues such as airports, stadiums, and corporate campuses as they seek to concentrate their efforts only on the most high-profile, profitable opportunities. The costs to install and operate the DAS networks are significant and typically cannot be justified by the wireless carriers from a return on investment perspective. The carriers may install and operate a DAS in stadiums and airports, but cannot justify it in smaller buildings.

Finally, another DAS variant is the large building which has an anchor tenant, often the building owner. The anchor tenant occupies most of the building but leases out unused space. The wireless carrier for the anchor tenant would probably bring in a single carrier but could benefit from the additional revenue for the leasers. Under the terms of most commercial real estate leases the building owner can pass through the cost of operating and maintaining the building to the tenants in the form of additional rent, *common area maintenance* (CAM) or utilities cost, including, but not limited to communication systems and other equipment used in common. If the DAS infrastructure is paid for by the DAS provider, and the wireless communication service is sold back to the building as a utility, a commercial lease would cover this charge as a tenant utility expense. In these cases the DAS provider should be a licensed *competitive local exchange carrier* (CLEC) with the transaction structured as a *utility easement in gross*, with a *utility service agreement* provided by the CLEC for a fixed term. In this scenario the DAS provider gets its return on investment and the building owner collects fees for utility charges since the tenants are the beneficiaries of the improved wireless service.

The business considerations for these models involve: (a) who owns the network, (b) who provides the capital, (c) agreements with each carrier, and (d) the service level agreements (SLA) for the DAS including their scope, quality, responsibilities and a contracted delivery time (of the service or performance). Part of ongoing service and maintenance includes changes to the building over time due to renovations and space modifications; such alternations could affect the performance of the building antennas. Also the latest technology changes (such as the recent shift from 3G to 4G) will impact updates and changes to the DAS deployment.

Below are some examples of real world DAS implementations:

▶ Merchandise Mart in Chicago: When it opened in 1930, the Merchandise Mart in Chicago was the largest building in the world, with 4 million square feet. The Merchandise Mart now has a distributed antenna system. The system is neutral host, allowing competing carriers to connect their radios and send out their signal. The DAS head-end or central distribution is in a 2,000-square-foot space, located on the 18th floor. Initially AT&T funded the installation. The leasing agents for the Merchandise Mart use the DAS as a sales point.

▶ Olympic Stadium in Baku, Azerbaijan: The DAS was installed to support the first European Games held at the 68,000-seat stadium in 2015. The DAS supports multiple radio frequencies for three mobile operators and ensures that capacity crowds at the Games received strong mobile services throughout the event.

▶ Gautrain Railway Systems: Gautrain Railway Systems is an 80-kilo-metre (50 mi) mass rapid transit railway system in Gauteng Province, South Africa, which links Johannesburg, Pretoria, Ekhuruleni and OR Tambo International Airport. A DAS was designed and installed in three underground Gautrain train stations. One of the challenges for the DAS deployment was that construction could not interrupt the train service. The master unit is located in one of the three train stations with the other two covered by remote units. The project supports both data and voice services for Vodacom and MTN subscribers in the Gautrain un-derground stations at Sandton, Rosebank and Park Stations. The next phase of the project addresses wireless coverage and capacity to the tunnels via leaky feeder and retrofitting the train cars.

11.2   Life Safety and Emergencies

A building with a DAS can increase the safety of the building's occupants. During an emergency the DAS will enable 911 calls within the building. The multiple antennas in a DAS will provide improved communication re-liability, and provide communications with First Responders, police, and fire fighters. Note that two national fire codes, the National Fire Protection Association's NFPA-1 and the International Fire Code were enacted in 2009 require in-building amplification systems such as a DAS. Such codes have been adopted in municipalities across the country.

The Federal Communications Commission (FCC) is responsible for en-suring that 911 services and other critical communications remain opera-tional during emergencies. This includes ensuring communications interop-erability among first responders and promoting use of enhanced 911 best practices. One of the main components is 911 call processing and delivery through *public safety answering points* (PSAP).

At the PSAP, the dispatcher verifies the caller's location, determines the nature of the emergency, and decides which emergency response teams should be notified. However, the use of cell phones makes determining a wireless 911 caller's location more complicated than determining a tradi-tional wireline 911 caller's location (where numbers are associated with a fixed address.) For callers placing wireless 911 calls, the FCC has required wireless service providers to make location information automatically available to public-safety answering point (PSAPs). Basic 911 rules require wireless service providers to transmit all 911 calls to a PSAP, regardless of whether the caller subscribes to the provider's service or not. The wireless service providers have to provide the PSAP with the telephone number of the originator of a wireless 911 call and the location of the cell site or base station transmitting the call. Eventually wireless service providers will be required to provide more precise location information to PSAP.

Emerging DAS systems can deploy a *self-organizing network* (SON). The self-organizing network can:

(a)   Automatically configure and integrate new equipment into the wireless network, something akin to plug and play. The network also discovers new components in a system without the need for a technician to manually reconfigure the equipment.

(b)   Automatically optimize the wireless network. It optimizes based on data from the system itself. An example of self-optimization is the automatic switch-off of a percent of base stations during the night which would reconfigure to cover a larger area or a significant increase in usage.

(c)   Self-heal, where the network can identify faults or failures in the network such as, base station failure. The network compensates and reconfigures to minimize the impact.

The DAS network can now move cell capacity from one location to another needing capacity, as well as identify what radio spectrums and functionality are needed on-demand in real-time. While all manner of base stations will be used in future network deployments, DAS will play a central role in providing on-demand capacity wherever and whenever it is needed.

While distributed antennas systems are primarily focused on large venues and buildings, there are smaller cells available for more modest spaces and buildings. These cells are called micro cells, pico cells, or femto cells. They are small mobile phone base stations that are connected to the wireless carrier's network via the Internet and typically used in areas where the mobile signal is weak. Generally these are very low-range, low-power base stations, able to be deployed in a home, home office, enterprise businesses, indoor or outdoor public space. Generally the equipment is provided by a wireless carrier mobile network operator, and operates in licensed frequency bands. A base station in a wireless carrier's network may have a radius of 2–30 km coverage; the small cells may have coverage radius of 10 meters.

12

Contents

DC Current

While AC current is the worldwide electrical distribution system for buildings, we are surrounded by electronic devices and equipment that operate internally on Direct Current (DC). These devices are plugged into a typical Alternating Current (AC) outlet, but the AC is then converted to DC, typically using a rectifier or convertor to operate the internal electrical components and equipment. A rectifier converts AC, which periodically reverses direction to DC, which flows in only one direction. This process is known as rectification. Now, many newly constructed buildings are deploying renewable energy sources such as solar or wind which can generate AC or DC power. One of the growing uses of DC is digital devices with transistors relying on direct current; smartphones, personal computers, flat screen TVs, and tablets. One example almost everyone can relate to is that there are 6.8 billion cell telephone subscriptions that require and consume DC power.

According to the director of the Power & Energy Initiative at the University of Pittsburgh, digital consumer devices and devices such as LEDs and solar panels account for up to 20% of total power consumption.

The history of AC and DC power started in the late 1880's and 1890's. Thomas Edison (who once held the world record of 1093 patents for inventions) developed the first commercial electric power transmission system which used direct current (DC). After he deployed about 200 power stations, a war of currents started. Opposing the use of DC was the inventor of transformers and alternating current, Nikola Tesla, and George Westinghouse, a proponent of AC power. Their basic argument was that DC power couldn't be transmitted very far, only around a mile and half, whereas AC current could be carried over hundreds of miles and was better suited for central power stations. The war of currents got nasty at times but eventually AC power won out for power generation, transmission, and distribution.

DC is described as the unidirectional flow of current; current only flows in one direction. Voltage and current can vary over time so long as the direction of flow does not change. Alternating current describes the flow of charge that changes direction periodically. As a result, the voltage level also reverses along with the current. AC is used to deliver power to houses, and office buildings.

Converting AC to DC or DC to AC generally results in small conversion losses. While the efficiencies of power conversion are dependent on current and voltage, conversion equipment is typically rated from 90 to 95% but can be much lower. So at best we're wasting 5–10% of energy in the conversion and more with multiple conversions using less efficient conversion equipment. To minimize conversion losses, the conversion of AC power should be upstream of individual devices.

For buildings to deploy DC systems would be a monumental change, but, the benefits of a DC infrastructure can be very persuasive. The following sections include observations and examples of existing or potential use of DC in buildings.

12.1 IT Networks

Most sizable IT networks have for years provided DC power to devices using Power over Ethernet (PoE). PoE is an IEEE standard where data communication and DC power are provided to a device over a four-pair Ethernet cable. Most buildings probably already have an IT network or infrastructure for DC distribution. The PoE devices typically include IP telephones, Wi-Fi access points, video surveillance cameras, access control card readers, and remote Ethernet hubs and switches. The use of PoE has gone beyond typical IT devices and been used in devices such as clocks, gas detection, and AV room controllers.

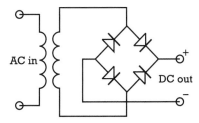

Figure 12.1 A rectifier is an electrical device that converts alternating currents (AC), which periodically reverse direction to direct current (DC), which flows in only one direction. This process is known as rectification.

12.2 Data Centers

Much of the equipment in a data center such as servers, motors, and batteries use DC power. The power conversions that takes place in a data center could include incoming AC power converting to DC at the UPS, converting back to AC, then finally converting back to DC within each server's power supply.

A few data centers have gone to a DC architecture where the main electrical AC feed into the data center is converted to DC and DC power is distributed. The benefits of this approach are threefold: (1) less power loss in multiple conversions of AC to DC and DC to AC, resulting in reduced energy consumption, (2) less space required by the DC infrastructure for equipment and the IT equipment can connect directly to backup batteries, and (3) a decrease in cooling requirements. Companies such as Facebook and SAP have piloted DC power in their data centers. Japanese telecom giant NTT has four data centers in the Tokyo region operating on DC; it completed a DC-based server center in Atsugi City, which is its first to serve external clients.

A Lawrence Berkley Laboratory (LBL) study indicated a 7% reduction of energy consumption and a 28% efficiency gain, comparing DC data centers to AC based data centers. The LBL study goes on to say "We were also able to conclusively demonstrate to the data center industry that DC delivery systems are viable, can be 20% or more efficient than current AC delivery systems, be more reliable, and potentially cost less in the long run."

Efficiency isn't the only benefit that DC power offers data centers. EMerge Alliance representative BJ Sonnenberg, manager of business development at Emerson Network Power, says that "DC data center power distribution equipment is generally more compact, takes up to 33 percent less floor space, and can be between 200 to 1000 percent more reliable" than its AC counterparts. A study conducted by Validus (ABB) and GE revealed that DC data centers should have up to a 36 percent lower lifetime cost.

The Electric Power Research Institute (EPRI) identified at least a dozen new data centers built in the last two years that operate on higher voltage DC, a figure that doesn't include the many central telecom offices currently operating on lower voltage DC power. The most powerful DC data center built to-date was recently announced by leading power and automation technology group ABB and Switzerland-based Green, an information and communications technology service provider. The data center is housed in Green's new Zurich-West data center expansion and is based on The Emerge Alliance proposed standard 380VDC technology. Up to a 20 percent reduction in power consumption is expected from grid-to-chip and in cooling.

12.3   Renewables, Electric Vehicles, Storage

The solar panels that have been installed in buildings generate DC or AC.

12.4   Lighting

Compact fluorescent lights with electronic ballasts and light-emitting diodes (LEDs) are 25% more efficient than traditional incandescent lighting. The benefits of LEDs include longer operating life, less maintenance and reduced energy consumption, something on the order of 13% for lighting. Use of DC power would further enhance the attractiveness of LEDs by eliminating the need and cost to convert AC power to DC as is currently done. Several lighting manufacturers have DC based LED systems

12.5   Appliances

Most everyday appliances and electronics operate internally on DC power. This includes computers, telephones, televisions, and coffee makers. Motors

Figure 12.2   Safety sign.

for applications such as refrigeration, ventilation or pumping that are DC-based are significantly more efficient than AC-based systems.

12.6   DC Power Infrastructure

Armstrong, a manufacturer of fl oors and ceilings has a product called DC FlexZone, a ceiling suspension system that provides an infrastructure for the delivery of low-voltage DC power based on conventional suspended ceiling installation practices and on a 24VDC Occupied Space Standard developed by the Emerge Alliance, a non-profi t industry organization.

12.7   Standards

The Emerge Alliance, an industry association with industry and research institute members is focused on the development of DC technologies and standards They have already developed a standard for commercial buildings (24 VDC), as well as a standard for data centers (380VDC), and will likely implement a standard for residential markets as well.

Building owners, contractors, and engineers will install and maintain traditional AC electrical distribution systems in buildings, possibly supplemented by solar or wind renewable energy which can generate DC or AC.

One issue with the adoption of DC is that while devices may be internally powered by DC they are generally manufactured and sold with AC plugs/power supplies. In new construction, this issue could be addressed by simply procuring new devices that can accommodate direct DC power. With existing buildings replacing current equipment with equipment that can handle direct DC is likely to be dependent on opportunities during renovations, procurement, and energy conservation strategies.

At times DC may face implementation issues. Relatively few people are experienced with the installation of low voltage direct current. Circuit

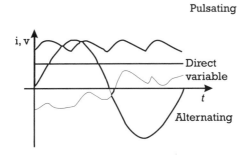

Figure 12.3

protectors, fuses, and insulation materials may need to be redesigned. Products and devices may need to be modified, training and test methods need to be developed, and adoption by architects, engineers, and contractors is critical.

Power Over Ethernet

Power Over Ethernet (POE) is probably the most under-valued technology in buildings. It provides both power and communications through one cable; and it can centrally monitor the devices. The devices can be telephones, wireless access points, cameras, paging speakers, card readers, lighting, and clocks.

Several major IT manufacturers have developed software to manage power to the devices, either turning the devices off and on, or dimming the power to the device, much like a lighting control systems. The POE management software essentially enables and disables POE ports on a network switch. The result is a reduction in peak energy demand for IT networks and the flexibility for network managers to set different power consumption for different IT devices at different times. The POE management software is typically a module in a larger suite of network management tools, with capabilities to scale from one network to an enterprise. Since the devices that are monitored and managed are already on an IT network, the monitoring of the devices is done with SNMP tools to monitor any device defined by a Management Information Base (MIB).

POE truly reduces the cost of building construction and operation. One cable can provide power and connectivity, thus eliminating the

Figure 13.1   ©Matt Buchanan, licensed under CC BY-SA 2.0

cost of materials and labor for one additional cable installation. Here's an overview of POE, its uses in products, devices, and systems that are installed in buildings, and the reasons why it needs to part of specifications for building technology systems.

13.1   POE Overview

PoE was created during the first wave of Voice-over-IP (VoIP) telephones and wireless access points (WAP) in the late 1990s. When VOIP and WAP were installed two cables were required, one telecommunications cable to connect to the data network, and another cable to a local power outlet. Older telephone systems that replaced VOIP telephones never required local power. Those older telephones were powered from a central telephone system, typically a private branch exchange (PBX). Many locations for VoIP telephones and wireless access points did not have local power available and power had to be installed at considerable costs. So the need for electric power for the first VoIP telephones and wireless access points undercut the attractiveness of those technologies.

That shortcoming of the VoIP technology spurred the effort to supply power to the telephones over the twisted pairs in the telecommunications network cable. The result was that in 2003 an IEEE standard was published allowing low voltage power, 48 VDC, to be transmitted over Category 3 and Category 5 Twisted Pair cable. With the current standard, IEEE 802.af, the maximum power that can be delivered to a powered Ethernet device is 15.4 watts. After counting losses, about 13 watts is the nominal power delivery available. Additional losses will occur with the use of switch mode power supplies.

Power is transmitted over the pairs of the Category 5 cable in one of two ways. For 1000BASE-T, which uses all four pair, cable pairs must carry both power and data. This method of powering is called *Phantom Power* (Fig.13.3). For 10BASE-T and 100BASE-TX Ethernet, which only use two out of the four cable pairs for data transmission, the power can be transmitted over the two idle pair. Transmitting the power over idle pair is called I (Fig. 13.4).

The device supplying power is called *power sourcing equipment* (PSE), and the device being powered, such as the VOIP telephone or wireless access point, is called the *powered devices* (PD). The PSE determines the method (galvanic injection or phantom power) used to power the PD. The PSE can either be a POE-enabled network switch (referred to as an endspan) or a device that injects power between a network switch not POE-enabled and the PD (referred to as a midspan). Midspans allow users without POE-enabled network switches to make their networks POE-enabled without procuring new network switches.

The initial standard is sufficient to power many low powered devices. However, the power rating is insufficient to power devices such as video surveillance cameras using pan, tilt and zoom (PTZ) capabilities, and video telephones. A second standard, IEEE 802, also known as *POE Plus*, is now being developed and will offer higher power levels, although still within the low voltage range. Expectations are that the standards for POE Plus will at least have a power rating of 30 watts, basically doubling the existing standard. One concern with the development of the standard is that higher power generates more heat on the cable pair(s) resulting in an increase in the attenuation of the cable. The result could be that cabling runs would need to be shortened to comply with the standards of TIA 568. Obviously, there is more to come on the development of the standard and its implications.

POE products started with VoIP telephones and wireless access points but have rapidly expanded into hundreds of certified products. These include surveillance cameras, clocks, intercom systems, paging systems, access control devices, time and attendance systems, and touch-screen flat panels. There's even a prototype POE electric guitar and a demonstration POE electric shaver for men. Here are a couple examples of POoE product types that improve the installation and functionality of systems:

▶ Access Card Reader: Network connected, POE powered access control card readers eliminate the need for a local AC/DC power supply and power the reader from the telecom equipment room, which typically is on an uninterruptible power supply.

Figure 13.2

▶ Paging Speakers: Each speaker is networkconnected and POE powered. The speakers become part of a VoIP telephone system, endpoints on the system much like the telephones. Thus a separate paging system, and interconnection of the paging system and telephone system is eliminated. In addition, the speakers do away with a limited number of paging zones and allow paging to specific speakers or predefined groups of speakers as needed. If your VoIP telephone system is networked between buildings or cities, so is the paging system and you can page from one building or city to the other.

▶ Clocks: POE clocks plug into the network and become end devices on the network much like VoIP telephones or card readers. The clocks can be automatically synchronized throughout a facility and managed by the network.

▶ Intercom Systems: Intercoms, which are used in classroom communications or as part of access control for doors and gates, are evolving to networked POE devices. As with many of the other POE devices, installation costs are lower and system flexibility is increased.

The benefits of PoE are numerous:

▶ At the top of the list is cost savings or cost avoidance. The cost of a power outlet includes conduit, wire, a back box for the outlet, and the labor of an electrician. If a power outlet does not have to be installed for the device, the cost of installation and construction is reduced. Purdue University installed over 1,100 PoE wireless access points across campus, and saved $350 to $1,000 per location by not having to install power. Others have estimated that an average cost to provide power to a device is about $864, while the cost of a POE network port is $47–$175.

▶ Increased Reliability: POE centralizes power distribution. Instead of a power outlet at each local device, power is now distributed from the telecom rooms, (a throwback to the older centralized telephone systems.) Centralized power makes it easier to provide uninterruptible and emergency power for critical hardware, thus increasing system reliability and uptime.

▶ System Management: POE allows the end device to be monitored and managed. Network switches provide management tools such as the simple network management protocol (SNMP), which allows staff to manage the end devices, including the power to the end device. You can remotely turn the device on or off or monitor the device's consumption of power.

▶ Move, Additions, and Changes: POE allows for slightly easier building renovations and rearranging of spaces since devices only need one cable. It's easier to install devices on walls or ceilings, and to setup temporary installations.

▶ International Applications: POE is being marketed and deployed worldwide, allowing manufacturers to avoid supplying different power cords for different countries and eliminating the need for installers to worry about power cords.

▶ Safer: When POE is used increased low voltage distribution is used to power devices, and less high voltage is used throughout a building. This results in a safer environment and lower power consumption.

In an environment where construction costs are steadily increasing, POE is a technology feature that can add long term value to a building. Over one hundred million network ports enabled by Power over Ethernet are currently shipping annually.

The initial POE deployments were providing 15 watts for VoIP telephones. The more recent advances of POE have included Cisco with 60w Universal POE (UPOE) technology driving the adoption of virtual desktop

Figure 13.3

infrastructure (VDI). The HDBaseT Alliance is promoting the POE HDBa-seT specification for consumer electronics, which can deliver up to 100w over twisted-pair cable, supporting full HD digital video, audio, 100BASE-T, and control signals in television and display applications. Finally, the IEEE is developing a standard called IEEE 802.3bt DTE Power via MDI over 4-Pair which will have remote powering applications and improved energy efficiency.

Microgrids

Centralized power plants have been around since around the 1880s. A hundred and thirty-five years later we're starting to see some growth in distributed generation of power at individual buildings, primarily through renewable sources such as solar panels and wind turbines. Between the distinct approaches of centralized power plants and energy generation at individual buildings is emerging something called microgrids. With microgrids the real estate developers, building owners, or the local community build the power grid for their large development, industrial park, campus or even an entire neighborhood. Microgrids are not new and are no longer just a concept. They are out of the experimental phase and are now commercialized with roughly around 300 microgrids operational worldwide.

14.1 Overview

Compared to a utility grid, microgrids are small or modest sized power generators which may include traditional fossil fuel generators, photovoltaic, wind, or fuel cells. Increased diversity of power generation improves the microgrid's reliability. The microgrid may be able to operate independently (remote villages or military

Figure 14.1

bases), or it could be connected to a larger utility power grid so that the microgrid appears as one customer or a provider to the larger grid. The organization and management of the microgrid could be a cooperative arrangement for a community, coordinated by developers, or may be a large campus with one owner.

Microgrids improve the reliability of the old grid and the overall power system. Locally generated power also lessens the burden on existing transmission lines and reduces energy loss in the transmission process. Typical microgrid transmission and distribution lines are short and generally any losses are negligible.

14.2   Potential Benefits

In time microgrids will impact each of us involved with designing, constructing, operating, and managing buildings. The key question is "Why would a developer or building owner be interested in a microgrid?" The rationale is persuasive:

▶ The microgrid improves power reliability. A microgrid with multiple generation sources provides diversity and therefore greater reliability.

Connecting a microgrid to the larger grid simply means increased power dependability.

▶ The microgrid has better potential to lower energy costs. While it's true that centralized power plants produce cheap power, there are opportunities to lower costs with a microgrid. For example, if a microgrid is connected to the larger grid the operator can use power from the larger grid when prices are cheaper than the microgrid; or conversely maximize the use of the microgrid when prices from the larger grid are high. Given variables such as time-of-day rates, demand charges, weather, potential demand response events, and load shedding scenarios, analytics can be applied to optimize when to use the larger grid or the microgrid. Eventually this will minimize the cost of energy, or could even facilitate making money by selling power into the larger grid. Owning a microgrid that is generating power offers more flexibility for the owners in managing their energy costs.

▶ The microgrid is energy efficient. A typical coal-fired power plant might be only around 38% efficient; meaning 62% of the original energy is not converted to electricity. Add to that another 7% loss in transmission and distribution. A microgrid with multiple generation sources is more likely to be more efficient through renewable sources, eliminating the transmission and distribution energy losses, and having the capability to recover and use heat locally. The result is higher energy efficiency and lower carbon production.

Research by Lawrence Berkeley National Laboratory sheds additional light on representing and calculating microgrid costs and benefits. Their case study ("A Framework for the Evaluation of the Cost and Benefits of Microgrids") covered a Canadian microgrid with 10MW peak load and a 6.2 MW average load. They looked at four metrics: the reduction of electricity purchased, investment deferral, the reduction of GHG emissions, and increases in reliability. They compared the microgrid to power from the larger

Figure 14.2

utility. For reduction of electricity purchase they found the cost of electricity via microgrid to be $5 less per MWh with additional benefits related to profits for the owners of the distributed generation in the microgrid (who are possibly the customers themselves) including sales of excess power to the grid. The load reduction provided benefits to the larger grid related to investment deferral; the larger utility company can defer or doesn't need to make a capital investment. There were also substantial benefits to society related to reduction of GHGs. Finally, the study monetizes benefits related to increased power reliability that accrued to customers, the microgrid operator, and the larger grid. The take-away from this study is that while most benefits from microgrids are related to the customer, everyone benefits.

14.3 Developers and Building Owners

A property served by a microgrid that provides more reliable power service at a lower cost adds value to the property. Studies have shown that tenants will pay slightly more for space that is LEED certified; the same may eventually be true for microgrids, maybe even more so because of the explicit benefits.

Building owners may also gain by deploying a basic microgrid and providing or charging for space in a microgrid co-location area for tenants to install their own generation equipment. This is similar to data center companies that sell space within their locations to multiple users.

There is now an international standard for microgrids reflecting the viability, credibility, interest, and momentum of this approach. The IEEE standard developed in 2011 (IEEE 1547.4) provides best practices for designing, operating, and integrating microgrid electric power systems. This includes the ability to separate from and reconnect to part of the larger utility grid while providing power to the microgrid. This standard addresses engineer-

Figure 14.3

ing concerns for microgrids specifically targeting reliability, contingencies, and interconnection requirements.

Pike Research anticipates the institutional/campus single owner microgrids will be the largest segment of growth with 53% of deployments by 2015, followed by commercial/industrial with multiple owners at 39% of deployments.

14.4   Macro versus Micro

One of the unsettled areas regarding microgrids is the role of the larger grid utilities, the legacy providers. It's generally assumed that microgrids will be deployed by non-utility developers, probably working for the real estate developer, building owners, or the neighborhood. These microgrid entrepreneurs and developers may offer improved power quality and reliability and tailor their services to specific customers.

Some utilities have opposed microgrids due to safety concerns; others support microgrids as long as the larger utility owns, operates, and bills customers, an approach that doesn't necessarily resonant with microgrid providers and building owners. Some utilities, such as the Sacramento Municipal Utility District have embraced the concept; SMUD is deploying microgrid architecture in their own corporate headquarters.

The potential utility grid versus microgrid conflict could resemlbe changes in the telecommunications industry in the 1980's and 1990's. That is when the telecom utilities were reorganized and decentralized, followed by radically changes due to technological advancements. The result being that their primairy business and largest revenue producer—telephone landlines for residences—evaporated with the onset of cellular and smartphones.

There is a significant trend to decentralize some energy generation. You see it in individual building uses of renewables and massive efforts to move towards net zero buildings. The deployments of microgrids are additional evidence of that accelerating trend.

14.5   Generating Revenue from Microgrids

While a few utilities may not support microgrids because they take a portion of the load away from the utility, other utilities may welcome microgrids to avoid large investments for repairs or upgrades in their transmission system.

The microgrid is an asset. The return on investment of the microgrid and the growth in its operating income is important to its financial success and wellbeing. Much like typical real estate transactions the location of the microgrid is important primarily because of particular state utility regulations and policies of the local utilities. Microgrids generally succeed in markets where electricity prices are high, and where there are large dense power users such as educational or business campuses or districts.

Many grid operators buy services from microgrids that are grid connected. Potential revenue stream for microgrid owners include:

▶ Buying or selling power into wholesale markets, within the rules and requirements of some grid system operators and regional transmission operators. Owners sell energy back to the utility grid for a profit once the selling price exceeds the cost of the microgrid's generation and storage.

▶ Demand response, where the grid operator will pay energy users such as the microgrid to curtail energy use.

▶ Capacity markets that have been set up to ensure that supply will be available when it is needed most. There is an incentive for microgrids to make their capacity available to electric markets where price signals alone would not work.

▶ Ancillary services that a microgrid can provide to grid operators. These are driven by short term unexpected disruptions in the grid's overall capacity possibly related to a faulty transmission line or an out of service power plant, or an intermittent renewable generation capacity like wind where resources are required to address inconsistencies in generation. A microgrid may address the short-term imbalances in electricity markets by dispatching resources.

▶ Cost savings by avoiding peak energy costs with self-generation and storage.

▶ Microgrid owners may also benefit with declining costs for solar panels and energy storage.

Microgrids and distributed energy are viewed as very attractive, because of reliability, flexibility, and profitability and will grow into an important role, providing reliable power at reasonable prices in a sustainable way.

15

Solar Energy

Imagine a building where the major components have photovoltaics embedded in the materials used in construction; the result being significant onsite production of solar power.

We're all familiar with solar panels on a building roof or a parking canopy. Solar panels have been the face of solar energy to date, although solar heat can provide energy as well. Integrating photovoltaics in a building is something completely different.

After decades of anticipation, the solar energy market has created a substantial and growing movement to integrate photovoltaics (PV) into buildings. This approach makes sense; solar energy within the building would generate power where it will be used and there is no need for any significant transmission or distribution infrastructure. This eliminates power losses and integrating solar power into buildings doesn't necessarily take up additional land or space.

Integrating solar cells into buildings focuses primarily on two aspects. One is the facade, which is essentially the exterior of the building. Facade systems include curtain walls (outer walls which are not structural), and spandrel panels (a wall between the head of a window and the sill of the window above in a buildings of two or more floors) and glazing. The use of integrated PV in the building skin replaces

147

Figure 15.1

conventional envelope materials, thus reducing the cost of the integrated PV.

The second aspect of integrated building solar energy is the roofing system. This includes tiles, shingles, and standing seam products for steel roofs and skylights. For example there are now solar shingles which look like traditional asphalt roof shingles, and metal roofing with upwards of 16% efficiency.

Besides solar facades or roofs there are innovative products such as walkable PV floors, transparent or colored PV glass, outdoor benches, and tables. Even the development of solar-powered concrete is underway.

The integration of photovoltaic into buildings may have started with the research and development of solar panel windows as a solar collector; this is now a reality and is in the marketplace. Wiring is embedded in the window frame and can provide direct current (DC) or be connected to a central power inverter to convert the direct current from the solar window to alternating current (AC) that is then fed into the electric panel for the building. This technology shows tremendous potential. Some of the current versions of photovoltaic windows can transmit more than 70% of the visible light, similar to tinted glass windows already in use in commercial buildings. The power conversion for the initial design of the windows was low but has steadily improved. One research team calculated that even with 5% efficiency these windows can generate over 25% of the energy needs of a building. Besides energy generation, the windows could also reduce infrared radiation, thus reducing thermal loads and operational costs.

The Whole Building Design Guide (www.wbdg.org) states: "PV specialists and innovative designers in Europe, Japan, and the U.S. are now exploring creative ways of incorporating solar electricity into their work. A whole new vernacular of 'Solar Electric Architecture' is beginning to emerge."

It's safe to say that integrating photovoltaics into buildings is innovative and will be disruptive for the traditional design and construction industry. However, if the approach provides beneficial results including lower energy and construction costs, greater utility, scalability, and creativity, building owners and contractors may see it as an opportunity and be attracted to its potential. Integrating photovoltaics into buildings will change the building design, with a clear priority of maximizing solar energy products and materials that can produce a substantial return on investment.

It is one thing to install a solar panel and quite another to construct a solar building. The design and construction of such a building will require some reeducation and training of engineers, architects and contractors, as well as possibly altering job responsibilities, trades, and skills. One would expect that substantial professional industry associations could assist in developing design guidelines and training.

Integrating solar into buildings makes sense for new construction where a building owner, architect, and engineer can design the integrated PV. This is referred to as *building-integrated photovoltaics* (BIPV). Existing buildings integrating photovoltaics are referred to as *building-applied photovoltaics* (BAPV) and are likely to be a more challenging undertaken. Ina new or existing building the architect, engineer, or contractor has to evaluate a proposed design related to solar access and identify potential use of photovoltaic systems.

Other aspects of designing photovoltaic into the building involve the building's location, its latitude, its structural aspects, nearby trees or buildings, shadowing, and average temperatures onsite. These factors must all be taken into account during the design stages where the goal is to achieve the highest possible value for the BIPV systems.

The majority of buildings using solar power are connected to a larger utility grid because the reality of using solar power for the entire building may not be possible. The building owner can operate the integrated solar power independently, but, connectivity with the grid provides a backup and could present an opportunity to sell power back to the utility. Both the building owner and the utility benefit with the grid being connected to BIPV. The on-site production of solar electricity typically coincides with the peak load times of the utility. The solar contribution reduces energy costs for the building owner while the exported solar electricity can help the utility grid during the time of its greatest demand.

The primary disadvantage of solar power is that it clearly cannot be created at night or during times of cloud cover. Solar panel energy output is maximized when the panel is directly facing the sun. This means that fixed locations have a reduced energy production when the sun is not at an optimal angle, unlike the solar farms that mount PV panels on towers that can track the sun to keep the panels at optimal angles throughout the day.

Solar cells convert about 20% of the sun's rays to electricity. While solar power can be a substantial initial investment, there is minimal maintenance, and after purchasing and installation it provides free energy. The

Figure 15.2

capital cost of solar power, batteries, and storage has continued to fall so that in many countries solar is cheaper than ordinary fossil fuel electricity from the grid. As the price of solar electricity continues to come down every year more and more countries will benefit from making the switch to solar when new capacity is added. The development of ultra-thin, lightweight, and highly flexible solar solutions is a key to the BIPV market.

China, Japan, and the United States have accounted for the majority of new solar energy capacity along with growth in Latin America, Africa, the Middle East and Europe, particularly Germany. In 2014 solar power accounted for more than 55% of new investment in renewable power and fuels. Industry analyst firm n-tech Research predicts the total market for building-integrated solar photovoltaic (BIPV) systems will grow from about $3 billion in 2015 to over $9 billion in 2019, then surge to $26 billion by 2022, as more truly integrated BIPV products emerge that are monolithically integrated and multifunctional.

As the cost of solar goes down subsidies will likely disappear. Success of BIPV will provide opportunities and major changes, and create new BIPV businesses. Manufacturers and construction companies will likely partner. The Whole Building Design Guide (WBDG) has suggested a need for a *solar*

Figure 15.3

energy architect, but, a team of systems integrators, construction firms, installers, manufacturers, and contractors will be needed.

Interestingly, some researchers think the commercialization of BIPV should primarily emphasize aesthetics of the materials and products. Different colors of PV windows would allow some distinct and artistic features, though some may want the products to look more traditional (roof shingles.) Solar panels on the roof of an old building can be an eye sore, so the improved aesthetics of BIPV may increase acceptance of the technology. With the PV being embedded in the materials and products no one really notices the PV. Smart aspects of BIPV could be automation related to energy, system integration, and building energy management systems.

Some important issues related to BIPV include the development of building codes and specific standards. Also to be considered are economic incentives from local, state or federal governments, optimal system orientations, the service life of the products and materials, their durability and capability of withstanding the weathering process, cost, and performance.

The amount of solar energy reaching the surface of the planet is so vast that in one year it is about twice as much as will ever be obtained from all of the Earth's non-renewable resources of coal, oil, natural gas, and mined uranium combined. Successful deployment of BIPV, combined with energy efficiency initiatives, can lead us to the goal of net zero buildings.

Wind Power

The earliest use of wind power was the sail boat. Ancient sailors came to understand the power of wind. Around 500 AD windmills were created to convert wind power into mechanical energy. The windmills liberated people from some manual labor and allowed them to pump and store water, as well as mill grain from their crops by turning stones.

Today's wind power is generated by modern wind turbines, a somewhat updated version of a windmill. Wind power is considered very clean and a major renewable energy source. Wind power does not generate carbon emissions, pollutants, or consume water.

The wind turbines convert the wind into electricity; the turbine blades are connected to an electro-magnetic generator that produces electricity when the blades spin. A typical small wind turbine may have several blades that are assembled atop a steel tube-shaped tower. Larger wind turbines may be over 300 feet and service a large development or campus. Multiple wind turbines in a location is called a wind farm.

There are some shortcomings for wind power. The largest issue is that wind speeds can vary throughout the day and year. However wind power may be used in conjunction with other electric power sources to provide a

Figure 16.1

reliable supply; this could include a connection to a utility grid, the use of other renewable sources such as solar, and power storage.

Can wind power be used in buildings? Yes, but there are many questions and serious concerns. Tall buildings are the best candidates because wind speed increases with height, but the wind flow is very turbulent on a tall building; wind turbines work better with smooth or regular wind flow.

If a turbine is installed on a building it's likely to be modest turbine, resulting in marginal energy production. In addition, noise produced by the rotor blades, along with the stresses and vibration of the turbines which can be transmitted to the building structure are all serious concerns. There are other issues with land use, aesthetic impacts, birds and bats having been killed (avian/bat mortality) by flying into the rotors, and obtaining insurance for the turbines.

Some manufacturers are using wind velocities from the building parapet or walls but the wind flow is very narrow thus minimizing energy production. One could install a wind turbine on a rooftop, although experts suggest that a wind turbine be elevated at least 30 feet within anything within 500 feet. In addition, rooftops with turbines would have to be quite sizable in order to be cost effective. Overall, the idea of a significant wind turbine on a building would have a very long return on investment, and provide numerous risks and challenges.

Large commercial wind projects require about 60 to 70 acres of land per megawatt (MW), primarily to facilitate and protect wind flow. A small amount of the land is for equipment, access roads and infrastructure such as conduit, cable, connectivity, and substations. Most of the land is a buffer zone to preserve wind flow.

Some farmers lease their farmland to a wind farm to provide another source of income. They can continue to farm the land although wind agreements can create complex legal and financial issues related to the land use. Wind-power leases often last 25–50 years. A frequent fear of landowners is that the developer or contractor of the wind farm will default and the land-owner will be left with large inoperable equipment on the property.

Wind speed is the critical feature of wind resources, because the energy in wind is proportional to wind speed. In order for a wind turbine to work efficiently, wind speeds usually must be above 12 to 14 miles per hour to generate electricity. Wind energy is very plentiful in many parts of the United States. Good wind resources have an average annual wind speed of at least 13 miles per hour.

Wind turbines are available in a variety of sizes. The largest turbines can produce enough electricity to power 1,400 homes. A small home-sized wind turbine can supply the power needs of an all-electric home. Single small turbines are used for homes, telecommunications equipment, or pumping water.

One economic issue for wind is that the cost of solar panels has decreased as the solar panels become more efficient. Solar power can now produce 60% more power for the same money as wind turbines; obviously effecting the value of wind turbines and the overall wind power market. One result of this market change is that the U.S. Department of Energy has ended their *Residential Small Wind Turbines* program. The cost of wind power has also been higher than conventional electricity generation, with a much slower return on investment. Although once the wind turbines are constructed, the cost for ongoing operations and maintenance is fairly low.

USA wind power topped 4 percent of the U.S. power grid in 2014 for the first time. In two states, Iowa and South Dakota, wind power now exceeds 25 percent of total electricity production. The United States is recognized as a world leader for wind energy production, primarily due to massive capital infusion via federal and state subsides, tax incentives and grants, that distort the real cost of wind power, with many foreign companies being eligible for the subsides. The tax credit for wind is $0.023/kWh.

Figure 16.2

The worldwide market for wind energy is growing. The growth year-on year is estimated to be 44%. 45% of new wind power installations in 2014 were in China, which had 110,000,000 homes powered by wind energy by the end of 2014 (total capacity 114,609 MW) (source: The Economist, 1 August 2015). In 2014, the United States represented 17.8% of the world's installed wind power. Twenty four countries have more than 1,000 MW of wind power installed across the world; 11 countries have installed more than 5,000 MW.

Integrated Building Management Systems

Almost every large building uses a building management system (BMS); it's the major platform for operating the building systems and the overall building. However, as we transition to more complex, higher performing, and energy efficient buildings, it is apparent that many of the traditional building management systems are not up to the task of monitoring and managing today's building operations. What are the shortcomings of the legacy BMS? The list is quite long but the major items include limited integration capabilities, inadequate and elementary analytic tools, proprietary programming languages, security, a dearth of software applications and legacy user interfaces.

17.1 Overview

To some extent, the BMS market has gotten to this point because of the business and financial aspects surrounding the procurement of a BMS. When a traditional BMS is sold and installed it's usually a small part of a much larger investment. The larger piece of business is the sale of building automation systems (BAS) controllers. It's the controllers' need for ongoing service, repair, parts and possible replacement over time that will generate significant recurring revenue for the equipment manufacturer or installation contractor. So the main building management tool, the one that provides the user interface for many of the building systems, often takes an inferior position to selling and installing the controller hardware. Some manufacturers may not put a lot of resources into developing a BMS product that will likely be only a very small part of a total sale.

Major BMS manufacturers have made some incremental improvements to their products. They may have added an *energy management software package*, reengineered an industrial automated process system for buildings, or even purchased smaller software companies thinking that would suffice. Despite these efforts,building management systems have fallen well short of where they need to be.

Part of the problem is that BMS manufacturers have not necessarily been good at IT and the BMS is an IT system: it's a computer server with a database, IP addresses and software applications, and connected to networks. What has developed at the industry level for building automation and IT is just a magnification of what is happening in many facility management and IT departments, that is, the readjustment of the roles of facility management and IT departments given the reality of the significant penetration of IT into building systems. The movement of BAS manufacturers into IT equipment, as well as IT companies into building controls has been an ongoing in the industry.

17.2   Escalated Complexity

The drive for improved building management systems reflects the increase in the complexity of new buildings. From an equipment or hardware perspective we now have buildings with energy and sustainability systemsthat were not commonplace even five years ago. These include systems like rain water harvesting, exterior motorized shading, water reclamation, renewable energy sources, electric switchable glass, and sun tracking systems.

Figure 17.1

Figure 17.2

Maintaining and optimizing each of these new systems is a challenge, further increasing complexity for building owners and facility management.

Another aspect of increased complexity is related to management decisions regarding building operations that now involve several other variables. For example, assume a building manager wants to respond to a *demand response* from the local utility grid, (a voluntary program that compensates retail customers for reducing their electricity use). requested by the utility during periods of high power prices or when the reliability of the grid is threatened. In making a decision on how to respond, the building manager has to take into account several financial and operational variables, including tangible and intangible benefits and cost. How much power load can I shed? How can I shed it? What's my typical demand profile during the time and duration of the event? How do I factor in the need to support the ongoing business? How do I implement, monitor, and measure? What's the effect on occupant comfort? How do I communicate to everyone affected by the event? Do I forgo the pricing signal to keep the business in normal operation? Do I use auxiliary energy generation? What's the maximum demand I can curtail?

These types of challenges are way beyond the typical question of "What should the HVAC set point be?" Obviously, some of these issues can be studied and a policy can be implemented by the building owner, but, any final decision would have to take into account real-time circumstances. This is where analytic and automation software tools of integrated building management systems can support the operations and facility personnel.

17.3 Specifications for the Future Building Management System (IBMS)

Some innovative medium-sized companies around the globe have made the first significant steps in providing building management systems that are beginning to meet today's challenges in building operations. What follows is a list of essential functionality of an IBMS:

The platform for the IBMS must be similar to that of smartphones and tablets. The base IBMS platform will have an operating system, much like laptops and smartphones where third parties provide the applications. That model is familiar and comfortable.

The base operating system for the IBMS will do the heavy lifting: acquiring data from different building systems, standardizing or normalizing the data into an open or standard database, possibly using something like XML/SOAP. This is extensive middleware, where the operating system can deal with the BAS communications protocol standards and data formats, as well as nonstandard data (i.e., some PLCs), with the BMS fully integrated into other facility management systems, such as work order systems, asset management, maintenance systems, inventory systems, and incorporating data from BIM files.

The IBMS must allow third-party applications for specific manufacturer equipment. Given that, every company that manufactures a valve, fan coil, sensor, or roof top unit will create an app for their equipment, much like they have for product objects in BIM. These apps are likely to be much richer in monitoring and managing the equipment and will create a burgeoning marketplace.

Third-party analytic software applications to optimize building performance are critical as they keep high performance buildings at peak performance and provide a rationale for similar analytics in many other building systems. Applications that can consolidate functions across systems, such as alarm management and master scheduling will become popular. Building managers will be able to test, compare, and choose the applications they need from a variety of third-parties.

The integration capabilities of the IBMS must be extensive. It has to go beyond typical fire, HVAC, access control, and elevator integration, and progressively integrate any building system, including the smart grid and external data such as weather and the financial metrics of energy markets.

The IBMS must be an open and secured system. It requires the tools that program the IBMS be transparent so that the building owner has choices in configuring, maintaining, and programming the IBMS. System security, which is almost nonexistent on traditional BMS, is a must on an open IBMS and probably best dealt with via IT security appliances and software.

The IBMS must be able to data mine and learn a person's use of the IBMS to identify their preferences along with data important to that user. Each dashboard is meant to convey important information and key indicators, and requires an examination of the needs of individual and group audiences. IBMS analysis of users' routines, usage, and interactions with

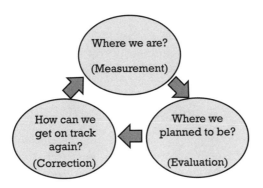

Figure 17.3

the IBMS will help to determine what the user needs to see, and possibly identify additional analytics and dashboards.

17.3.1 The Benefits of an IBMS

Some of the major benefits of an IBMS include:

▶ One Management Platform: Large commercial or educational campuses have many buildings and some have a separate BMS for different groups of buildings. Multiple BMSs on these campuses results in inefficiency; facility engineers have to go to multiple locations to monitor and manage the building systems. One management platform with internet access allows the facility engineer to view any BMS data anywhere, via tablets or smartphones.

▶ Consolidated Data: A single platform with one consolidated database has a number of benefits: It provides central control of the data, greater access to information, the ability to improve decision making, and efficient and effective use of resources. It also facilities facility engineers to quickly identify potential analytic applications. Data is consolidated onto a single system to improve reporting, information management, and decision-making. Integrating and managing applications from a single workstation allows facility-wide insight and control for better performance.

▶ Improved Security: Overall, the security vulnerabilities of traditional BMS systems have not been adequately addressed. Security is likely to increase through consolidation on one platform because there are fewer endpoints for attack. Imagine someone remotely setting off the fire alarm, opening doors, turning off the HVAC controls, accessing a video surveillance camera or turning off all the lights in a building. All of these are possible life safety or major business disruption issues. The IBMS can provide improved security consistent with the best practices of IT management.

▶ Increased Operational Savings: The IBMS is more efficient and effective. The result is reduced operational costs, a tool to support facility staff, and streamlining the process to training facility operators.

▶ Analytic applications: Most BMS systems provide data on building equipment and it's up to an engineer to analyze and interpret the data. Regardless of how talented or knowledgeable the engineer may be it is better if a software application can support the engineer in the analysis. That is why one of the emerging software applications for large HVAC systems is fault detection and diagnostics (FDD) and predictive analytics. These tools generally support the optimization of the HVAC system and can result in significant energy and cost savings. Very few BMS systems have such sophisticated tools while most of the IBMS systems do.

▶ Energy efficient: Real-time view into facility operations and deep trend analysis provide data-driven insight to optimize your energy management strategies and minimize operational costs.

▶ Flexibility to grow and expand: The powerful combination of open systems protocols and a scalable platform means the IBMS can help support growth and expansion of the system in the future, from one building, to a campus or world-wide real estate assets.

▶ Data is consolidated onto a single system to improve reporting, information management, and decision-making. Integrating and managing the HVAC, energy, security, digital video and life safety applications from a single workstation allows facility-wide insight and control for better performance.

▶ Integration Capabilities: Buildings are more than just HVAC, fire and security systems. Most IBMS systems provide integration capabilities for many other systems. Some of these offerings will be for other building systems such as lighting, window shading, and power management. Others will be business systems such as accounting or human resources. Other information relevant to building management may be outside the organization, such as weather data. Overall, the building manager needs data from several sources in order to create information and this requires integration, not silos of data. Improved or advanced IBMS systems will need to have middleware software tools to be able to standarde data from a variety of sources and systems into an open database structure.

▶ Integration of systems not only involves obtaining data from multiple sources to create information, it also means deriving more functionality from the systems working together. That means a sequence of operations between two or more systems, where an event or condition of one system can trigger or initiate actions by other systems. BMS systems generally do integration between fire alarm, HVAC systems and access control, but, the opportunity for other similar coordination exists. Some integration such as aligning building occupancy with energy related systems can have significant payback opportunities.

Future building management systems will reinvent a lethargic industry. It's also likely to spawn new companies and manufacturers, provide more choices for users and the buyers of such products, and do so at lower costs.

Contents

Dashboards

The question is not what you look at, but what you see.
—Henry David Thoreau

18.1 Overview

Buildings can generate a lot of data. Much of the data is collected through building automation systems and their data points, sensors, meters, databases, and measurements. Each of the building systems provide data: HVAC, power management, access control, lighting, life safety, solar panels, structural monitoring, personnel RFID systems, motorized shades, parking guidance systems, and electrical switchable glass.

However, lots of data does not necessarily mean lots of actionable information. Data is raw material. Its real value is being transformed into useful information where some intelligence has been gleaned from analyzing or studying the raw product. The final link in this chain, and probably the most important, is the user interface (UI) or the human-machine interface (HMI), where the actionable information is presented to the person who will act upon the information. For years the user interface was BAS graphics from a building management system typically used repetitively from client to client and generally had few improvements. Today more advanced integrated building management systems use visualization software. This software can take abstract data and develop images that can aid a user in understanding the meaning of the data. Many of the recent building management systems can incorporate popular browser-based dashboards to present information to users. This allows facility management personnel to remotely access data using smartphones and tablets and to present information to users.

What follows are some suggestions and guidelines on creating dashboards for facility and energy management covering what information is needed, how that information should be presented to a user, and a couple of industry examples.

18.2 What to Present

Dashboards can provide relevant and timely information to several organizational levels or groups involved with a building's performance. These different users can be facility technicians, managers, C-level executives, tenants, occupants or visitors through kiosks or a web page. The information provided may cover the specifics of particular building systems such as HVAC, electrical or specialty systems, but. they tend to focus on energy usage, costs, KPIs, trends, alarm management, and comparisons with similar buildings. So the first and most important step ise determining the right information for the intended viewer of the dashboard.

Facility technicians have different information needs than C-level executives or the general public. For example, a facility engineer may be interested in subsystem alarms and alarm management. In this case the dashboard needs to display alarm priority, escalation status, alarm acknowledgment, repetitive alarms, out-of-service alarms and subsystem communications or component failure. C-level executives, such as directors of facilities, sustainability, or procurement may want information on energy usage and cost. In this case the dashboard should display the usage and costs of a building's comprehensive and individual utilities, budgeted versus actual utility costs, budget deviations, comparisons with other similar buildings, and meter output for alternative energy sources such as photovoltaic and wind energy.

In developing a series of dashboards, you need to identify what insight each user or group hopes to gain by using the dashboard, and what information at what time interval is needed to support their decision process.

Dashboards will be fed from data and that data will probably need to be collected from several sources: building automation systems, specialty systems, and business systems. For example, if you're creating energy dashboard, energy usage may be generated in a BAS, whereas the cost of the energy may be in a database in the company's accounts payable system, or real-time data from the local utility. If you're a healthcare organization you may be interested in metrics such as energy use of an MRI machine per patient, and need patient counts from business systems. If you are a retail company it may be energy use per customer or per sale and you need customer and sale data from the business systems.

To gather all the information needed for a dashboard you may need a middleware platform to normalize and standardize data generated from several sources in different database formats. This would allow a flexible and consistent platform for the dashboard, but, could potentially trigger additional data management with large amounts of data. Dashboards are typically used for high-level performance summaries. Some dashboards such as analytical dashboards need to drill down to specific data, so data management can depend on the specific use of the dashboards.

18.3 How to Present the Information

Dashboards are meant to convey essential information quickly and clearly on one screen. Most importantly they do so based on their visual design. Visual design is much more than nice graphs and spreadsheets. It involves how human beings perceive and act upon visual information, a science in the realm of human factors and cognitive psychology. Although this may sound like you'll need a PhD to understand it, it actually is somewhat intuitive.

It all starts with something known as *preattentive variables*. These are the attributes of the dashboard that humans subconsciously pay attention to before they consciously know they are paying attention (thus preattentive attributes.) This innate perceptual and cognitive capacity to pay attention unconsciously evolved in human centuries ago. So if you're designing a dashboard to quickly display information, you take into account the preattentive variables to get the user's attention before they know they're paying attention.

18.3.1 The Position of the Information on the Dashboard

Information can be emphasized or deemphasized by its position on a display. The visual dominance is the center of the screen. Depending how the culture reads (left to right, or right to left) the other area of dominance will either be the top left or top right of the screen. The other corners are neutral, or in the case of the bottom right, actually deemphasized. So the most important data, such as key performance indicators, has to go in the center or the top left of the dashboard. This is especially true if secondary data on the dashboard can only be understood after an understanding of the most important information.

18.3.2 Color

Color is another preattentive variable that can aid in the clarity and quickness of understanding information. Here's an example of how easy it is to pick out data based on the blue color. In fact, if there were many more data points, the time needed to scan and quickly pick out the blue data points would be about the same.

Our perception of color is relative and dependent on the context that surrounds it, so selecting the color of the object and a contrasting color for the background is important. There are variations of the use color as a preattentive means, such as color hues, brightness, and color saturation.

18.3.3 Shapes and Sizes

Shapes are also a preattentive variable that, like color, can assist the user in quickly differentiating data sets. The size of a shape may be use to convey quantities or magnitude. Enclosing a data set in a border or using icons to provide meaning or focus attention are also positive uses of forms and shapes.

The reason preattentive attributes are important is simple. Dashboards should quickly and instantly allow the viewer to grasp the information important to the user. Preattentive features are just a head-start on that process, providing information to the viewer before the viewer consciously knows he or she is paying attention.

18.4   Industry Examples

Here are a few industry examples of well executed building and energy dashboards:

▶ National Research Energy Laboratory (NREL): This is a Demand Response dashboard Enernoc created. It's one screen with the most important information in the primary screen position. Viewers can also interact with the dashboard to calculate and change timelines.

▶ Lucid Design: Lucid is best known for their work in higher education and this dashboard addresses electrical use in a dormitory. Note the positioning, the colors, the user options, and the clarity.

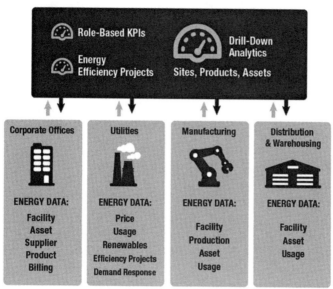

Figure 18.1

▶ Controlco: This dashboard is meant for a building engineer to analyze a system, in this case a chiller plant. It quickly conveys the system data points, alarm status and provides options on the left for further information.

Dashboards are the end result of a lot of work to identify, gather, and standardize data and to clearly understand the roles and the needs of people involved with facilities and energy. If they succeed in providing actionable information in a timely fashion, dashboards have a positive effect on managing a building's performance and operation.

Contents

Video Surveillance Systems

Video surveillance systems are used to deter crime and identify criminals when a crime has been committed. A video surveillance system can monitor stores and stock, provide a visible presence that video cameras are used in a building, allow building management to see what is happening at any time of the day, identify exact times when crimes have been committed, provide an identification method by which people can be screened before entering a building, and allow security personnel to check who is in a building at any given time. A complete video surveillance system consists of cameras, a control station, servers, hardware, operator work stations, software, cable, infrastructure, and junction boxes.

Video surveillance systems should be integrated with intrusion detection, access control, and electronic personal protection systems. The surveillance system may be comprised of both fixed and *pan, tilt and zoom* (PTZ) high resolution color cameras for monitoring the perimeter and interior of the building.

When we think of analytics related to building systems we generally think of predictive analysis or fault detection and diagnostic software tools related to HVAC systems. Video

Figure 19.1

analytics, that is software that can analyze and identify people, objects, and events in many ways can be just as important in providing information on building use and performance. Video cameras can be multifunctional. The analysis of digital images addresses physical security, but, goes way beyond that to provide data and information for building life safety, energy management, and overall building performance. This one device, the video camera, has a variety of uses for sensing and gathering data about the building condition and performance. This is a good thing, as more high quality and relevant building data is critical in generating actionable information and is a key to better building management and performance.

If you assume that the video camera is an extension of the human eye, the analytical software is the extension of the human brain. Cameras currently available can detect smoke or fire, identify specific people, detect motion, determine if objects have been moved, and provide occupancy data including the actual number of people in a space. Generally, if you can develop a pixel template of the event or condition you are trying to track, the video analytic software can detect the event or condition.

The video analytic process starts with cameras capturing successive digital video images of a coverage area. The digital image consists of pixels, a contraction of the words picture element and the smallest element of the

Figure 19.2

digital image. The analytic software first analyzes pixels, their patterns, the adjacency of pixels, the changing of pixels over time, and then compares the pixels to a database of templates of objects, conditions, and events. When the software gets a reliable match between the digital image of the coverage area and its database of templates or conditions, the video system identifies an event, state, or situation.

Video cameras are a staple of physical security systems. In the past, you typically had a security operations center where personnel viewed the feeds from the cameras and subjectively determined whether an event or action had taken place that warranted action. One of the largest benefits of using analytics in a typical video surveillance security system is improved detection and identification of threats, conditions, and events (machines outperforming humans). The software is working 24/7 with a constant level of accuracy. Also many video surveillance operations are not real time, with video simply being archived, available for search and review after an incident. Even if the system is manned, the attention span of personnel in a security operations center is oftentimes very short and inconsistent.

The analytic tools related to video cameras are extensive. As one would expect, most are geared towards some aspect of security and they include:

▸ Facial recognition: Video cameras can be used for recognizing people who are then given access to a building or space.The camera can recognize people that may be threats.

▸ Motion detection: The analytics can detect motion within a camera's coverage area, triggering an alert.

▸ Missing objects: By comparing digital images the analytics can detect if objects are missing or if a new object has been placed in the coverage area.

▸ Reading license plates: The cameras can read license plates to determine if particular vehicles have access to parking garage or building.

Some of the more innovative and interesting aspects of video analytics are people counting for occupancy and using video as a detector of fire and smoke. The following section contains an overview of these two applications.

19.1   Occupancy, People, Counting and Energy

The simple act of counting people entering or exiting buildings can provide very valuable data which can be used in a number of different ways. One

of thee primary uses is for energy management. At the core of building
energy management is an alignment between energy consumption and occupancy of building space. Getting data on energy consumption is fairly
easy through the use of meters or utility billings. Obtaining data on accurate
occupancy is much more challenging. Aside from retail buildings, very few
building owners or facility managers have data on the number of people in
their buildings, a time profile of their occupancy, or the count of occupants
within the building

   The general options for gathering occupancy data are infrared sensors
around the door frames, people carrying RFID tags, or access control card
swiping, all of which have potential shortcomings.

   In a video analytic solution cameras are placed above an entrance or
exit that can detect people and their movement in or out of the building.
Systems typically collect statistics on space occupancy and variations of occupancy during the day or by day.

   One example of the benefits of occupancy and people counts is using
actual people counts at the beginning of the workday to startup and ramp
up the HVAC system properly. People counting can also be utilized for the
ventilation of certain spaces. For example, one of the advanced HVAC control approaches is CO_2 *demand control ventilation* (DCV). Thi is best used in
large areas, open office spaces, theaters, assembly areas, and ballrooms. A
CO_2 sensor is used to optimize outdoor air use along with the energy required to condition the air. The CO_2 sensor is really a people counter or at
least a metric that helps reflect occupancy. However detecting occupancy
through CO_2 sensors has its limits, can be unreliable, and provides poor
estimates of occupancy. People counting technology with accuracy rates of
95% provides more reliable and accurate estimates of occupancy. Not only
can you use the occupancy data to improve energy demand, but, the occupancy data can be used for to evaluate space utilization.

19.2 Video Smoke Detectors

In the life safety area, video analytics capture images and use an algorithm
to compare the images to a database of smoke and fire patterns. Typically
these tools are assessing changes in brightness, contrast, motion, and color.
The use of video in this manner has several advantages. The cameras may
reduce or eliminate the need for traditional smoke detectors. Also, you can
use the video smoke detector in spaces where a traditional smoke detector
may not work, such as vehicle tunnels, high ceilings or in locations where
the detection device may be exposed to outdoor elements. The first recognition of video images used for fire and smoke detection was in the 2007 edi-

Figure 19.3

tion of NFPA 72. (As always, their use should be discussed and approved by the local authority having jurisdiction (AHJ), generally the Fire Marshall.)

With typical smoke detectors the smoke from a fire has to move to the smoke detector causing *transport delay*, essentially wasting time to trigger the detector. Video smoke detectors have no such delay and therefore are quicker resulting in less damage and threat to life. When a fire occurs, minimizing detection latency is crucial to reduce damage and save lives.

While the main purpose of video cameras is physical security, analytic software allows for more enhanced applications. In the future we can expect video cameras to take on the role of building sensors, not only in calculating occupancy, but, sensing other characteristics such as light levels or even thermal comfort. Video cameras are the Swiss Army Knife (probably the most famous name in multi-tools) of building sensors.

Access Control System

Access control systems are a critical component in smart buildings as security has become more important. The access control system is essential for life safety and is interfaced to the fire alarm system to facilitate building egress during life safety evacuations. Access control systems must integrate with several other smart building systems (video surveillance, HVAC, lighting, and others) as well as share data with business systems, such as human resources, time, and attendance. In a smart building, one electronic access control system for nonpublic areas should be deployed. Within secured areas the access control system would provide two levels of authentication. The system should support offline operation to allow doors to function if network connectivity is lost. The access control system should be supplemented by an intrusion detection system at potential unauthorized entrances, such as windows.

   The initial access control systems were more mechanical than electronic; for example, the use of the familiar pushbuttons on a door lock that required some type of combination to be unlocked. Access Control systems are used in a variety of markets and diverse applications. Some examples include:

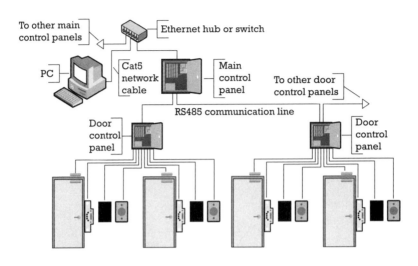

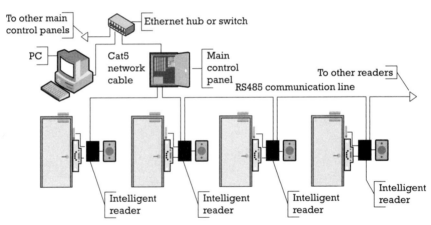

Figure 20.1

▶ Card access control for building entrances and exits;

▶ RFID for inventory and asset tracking;

▶ Pharmaceutical production in order to meet FDA requirements;

▶ Tracking of inventory and containers at seaports;

▶ Employee entrances for time card applications.

The access control system should be configured to maximize security. For example, its use of vertical transport systems (elevators) can be based on occupant identity provide selective access to floors, as well as spaces such as parking garages. Security levels should be determined by individual, floor, area, and access privileges can be changed in response to building occupancy states (i.e., time of day). An access control system can also generate anonymous occupancy statistics for building spaces and zones. Such data can be used to correlate occupancy to other building systems such as energy consumption or lighting schedules. While access cards are generally used in many systems, biometric authentication may be utilized for an additional degree of security.

One of the largest problems with access control systems can be *piggybacking* and *tailgating*. Piggybacking happens when someone with legitimate access to a building allows someone without access to the building to enter with them. Tailgating involves taking advantage of someone who legitimately has access to the building, where a trespasser enters the building with a person (or group of people) without their knowledge. One way to prevent piggybacking and tailgating is a *mantrap shield* that uses sensors to ensure that only one person is entering the building using one set of credentials. Mantrap shields can also be configured with separate compartments so that if more than one person is sensed passing through the first door, the second door will remain locked.

Access systems have leveraged the IT infrastructure, which eliminates the need for local AC power by using Power over Ethernet (POE). This consolidates the sytem, saves labor cost for cable installation, reduces the time to install system devices, and provides a large base of management tools and support.

The headend of the access control system uses IT infrastructure to interconnect and share information with a databases residing on other networks. The system headend and the door controllers house data regarding access credentials. Credentials consist of who gets access to what spaces during what hours or days. This data needs to be shared, and interacts with databases residing elsewhere such as Human Resources or Student Records. Those databases have now moved to open standards for databases such as ODBC, SQL, and XML to ease the integration. Data exchange rules between access control and other systems would use standard protocols and address all physical, virtual and calculated points, and operating parameters.

The access control system monitors and controls facility access via electronic access controllers utilizing card readers, keypads, and biometric devices. The system can monitor alarm points, control output devices, manage elevator control floors, and maintain an operator audit trail of operator activity along with access control and alarm activity. The system would typically focus on electronic door hardware and integrated devices, but, can also manage motorized access gates and associated devices.

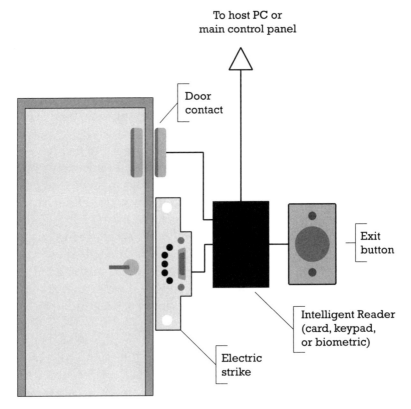

To host PC or
main control panel

Door
contact

Exit
button

Intelligent Reader
(card, keypad,
or biometric)

Electric
strike

Figure 20.2

Access control is designed to be activated by an authorized card receiving a signal from a reader and activate an electronic opening device (e.g., electric lock, or gate) to allow access. Should an attempt be made to enter this system with an unauthorized card, the electronic device is not activated, thus denying entry. This system is also designed to provide for an override by the operator locally or remotely to activate electronic door devices, allowing access for circumstances not normally programmed into the system.

There are codes and standards that cover the installation, performance, testing, and maintenance of access control systems. These include:

▶ National Fire Protection Association (NFPA) 731;

▶ National Electrical Code (NEC) and applicable local Electrical Codes;

▶ Underwriters Laboratories (UL);

▶ American National Standards Institute (ANSI);

Figure 20.3

▶ Federal Communications Commission (FCC) including Part 15, Radio Frequency Devices.

The management and monitoring of access control subsystems uses a server with software that provides the interface with the field-installed controllers along with the ability to manage the access control database. The database consists of the records that determine access privileges, field-installed controller behavior characteristics, history transactions, and third-party interface data.

One of the key component of access control system is the *access control unit* (ACU), which is an intelligent peripheral control unit that provides the interface between the management and monitoring subsystem and the devices installed at the access portal for the purpose of restricting access and monitoring the portal status. The card reader or a biometric reader captures the credential information and passes it to the ACU for processing.

Many access control cards use the patented Wiegand technology. Wiegand is the trade name for a technology used in card readers and sensors, particularly for access control applications. Wiegand devices were originally developed by HID Corporation. A Wiegand card looks like a credit card and uses a patented magnetic principle that uses specially treated wires embedded in the credential card.

The access control system monitors and controls facility access via electronic access controllers utilizing card readers and keypad devices. The system shall be capable of monitoring alarm points, controlling output devices, and managing elevator control floors. The system shall maintain an operator audit trail of operator activity and all access control and alarm activity. Each controller shall be able to manage the hardware necessary to secure one door and maintain a full database of information so that access decisions are made within seconds. Interval data shall be collected at a maximum of 15 minutes every day per year.

In the following sections we will dicuss four fundamental peripheral components in an access control system related to the doors.

20.1 Door Contacts

Door contacts are electromagnetic components that monitor the closed/open status of the door being controlled. These components consist of a pair of electromagnets. Typically these electromagnets are located at the top of the door with the passive magnet installed on the door and the electrified magnet installed on the doorframe. The magnets align with each other creating a magnetic field between them. When the door is closed, the magnets are aligned and the magnetic field is undisturbed. This position notifies the system that the door is closed. When the door is ajar, the magnetic field is disturbed and the system is notified that the door is unsecured.

20.2 Request-to-Exit

Request-to-exit (REX) monitors activity immediately inside of a controlled door sensing that someone is approaching the controlled door from the inside and allows them to exit without activating an alarm. The door controller has an input for each door for the REX. When the REX push button is tripped, the door is then open or unlocked.

20.3 Electrified Door Hardware

Electrified Door Hardware allows the door to be automatically locked and unlocked by the EAC system. These components can be contained in the door or the doorframe. For example, an electric strike will be located in the doorframe and an electrified panic bar will be located in the door. Determining the best method requires an understanding of basic door hardware and functionality.

20.4 Readers

Readers are the devices that interface with personnel desiring access to a facility. There are several different types of readers. Some common types are Contact Smart Card Readers, Proximity Readers, Swipe Card Readers, Bar Code Readers, Insertion Readers, Biometrics Readers, and Key Pads. For additional levels or security, it is common for readers to contain a keypad in addition to one type of card reader.

Proximity Readers are a contactless technology that utilizes RFID technology operating in the 125 kHz operating frequency. MIFARE Readers are also contactless technology that operates at the 13.56 MHz operating frequency. MIFARE technology is primarily used for e-purse applications such as vending, fare collection, transit, prepaid metering, phone cards, and toll roads.

Access control is a fundamental security concept. Without control of access, there is no security. The key is effective identification and authentication of people, via fingerprints, smartcards, and encryption keys and to have actions in place to make sure the identity isn't being used by the wrong person.

Given the larger concern for security, its likely advanced technology of access control systems will be vastly improved, and access control will integrate with emerging technology such facial recognition, eye tracking, and indoor positioning.

Contents

Maintaining High Performance Control Systems

We rely on control systems to monitor and manage our building systems. For the most part it's been assumed that once the control system is installed and configured it will work for years with little attention and minimal maintenance. Some systems may be trouble-free, but, the majority of systems will need regular attention and maintenance. Over time hardware will fail, software parameters change, and the control system will drift from its original configuration and performance.

The role of control systems is somewhat undervalued. When you examine the most complex system in most buildings, the HVAC infrastructure, you find that it's the HVAC control system, not the HVAC equipment, which produces the most operational issues and is the leading cause of inefficient energy use. Lawrence Berkley National Laboratories examined 60 buildings and found the highest frequency of common problems with HVAC was in the control system. Texas A&M University research determined that of the operational and maintenance measures that could produce significant energy savings, 77% of the savings were from correcting control problems. Maintaining a high performing control system involves regular maintenance, up to date software, data management, and organizational policies. Issues that cause problems with a building control system are the same challenges all of us have had with our computer or smartphone: problems related to software, hardware, communications networking, and user mistakes. What follows is an overview of some typical control system issues along with recommendations on how to keep the system performing at a high level.

21.1 Software Issues

Software is probably the number one issue with control systems. Given that control systems are networks monitoring and managing data points and running control sequence programs, problems with software and data management is not a big surprise. Problems can crop up with the initial configuration of the data points in new or replacement BMS systems. In existing buildings you may have multiple naming conventions, a lack of as-built control drawings, and overall poor data management, making it difficult and time consuming to obtain accurate information on point configuration. Even if you get accurate information on the data points, there may be human errors in configuring the points in the software.

Beyond the configuration of individual points is the management of the control strategy software where both the control logic and appropriate parameters must be identified. If the control logic between different HVAC equipment is not sound, parameters for set points or ranges for other data variables are not suitable, or the space use has changed, you have a control system that is providing suboptimal performance for the underlying building system.

The building management system (BMS) that manages and monitors controllers, data points, and control sequences can also be a software issue. Many of the problems are related to the BMS being an IT device. It has databases, operating systems, software applications, requirements for security, and a need for IT support. With no underlying support from IT or a lack of IT expertise within Facility Management, software problems will occur. In addition, a typical BMS system also has problems of omission. The BMS may not have intuitive graphics, analytic software or any application tools to support technicians and engineers in quickly identifying problems and solutions.

21.2   Communications Issues

Assuming that the software and hardware of a control system is properly working, network communications problems will usually involve cabling faults, improper cabling installation, excessive network traffic or the interface into IT network equipment. Cabling can get damaged and network connections can become loose resulting in a loss of signal across a communications span. If you're using wireless technology you can lose contact when using an unlicensed frequency and when other equipment using the same frequency is introduced into the space, causing interference.

The interface of a control system into the client's IT network is another potential source of communication issues. This involves cabling into an IT network switch, and possibly additional equipment such as a gateway that may be need to translate the control systems protocol and data format into an acceptable format for the IT network.

With field controllers where the controller uses an analog signal to communicate with the field device, such as a temperature sensor, the issue is calibration. A typical sensor may signal output via a range over a DC current (4mA to 20mA), identifying the zero and maximum level of the output of the device. These analog communication links need to be calibrated, configured, and validated to ensure the controller is getting accurate data.

21.3   Hardware Issues

At some point hardware devices fail, so every piece of hardware in a control system is a potential point of failure and potential problem. At the lowest control system level we have devices that provide for communication of the monitoring data to the system, with the data usually being a measurement

Figure 21.1

or state of a device. These are the typical sensors, relays, and transducers. In addition to complete failure of a device (such as a sensor) you can have an operating sensor that's just inaccurate. Sensors need to be recalibrated on a regular basis (many organizations calibrate on a regular schedule.) The issue here is that the control system may be receiving and acting on poor quality data.

At this same control system level are devices that the control system is managing and controlling. These are devices such as valve or damper operators and variable speed drives. Failure of the device, such as a leaking control valve, really negates the control request and overall control strategy of the building system.

The controller themselves may fail. This is typically related to the controller's circuit board, either the components on the circuit board or the board's ability to bond different parts of the board.

21.4   Operator Issues

Operator issues are the human aspect of control systems. A typical example would be an engineer or a technician overriding a control parameter such as a set point without documenting the change. The override affects the control system, as well as other engineers or technicians that may be working on that portion of the system, but, not informed of the change. This human aspect of the control systems plays a part at the larger organizational level of a the facilities management department, where the operation doesn't emphasize preventative maintenance, training for its staff, or maintenance of the control systems.

21.5   Steps to Take

▶ Inventory and document your control systems. Identify the location of all equipment and the version of the components and software.

▶ Recalibrate your sensors as well as the analog signals to the field controllers.

▶ Gather and manage data related to the control systems such as as-built control drawings and points list. Don't wait for an emergency to have to scramble to find everything.

▶ Audit and evaluate the existing controllers for parts availability, service, and overall capability.

▶ Develop a step-by-step methodology for troubleshooting. For example, you may start with the information from the BMS, check the controllers and any IT network involved, which should help in localizing the problem. After that you may need some instrumentation to check cables, communications signals, and voltage or current between a field controller and a sensor or actuator.

▶ Assess the needs of the building owner and operators. If you are dealing with a portfolio of buildings, get a BMS system that can provide an enterprise-wide solution rather than managing buildings individually.

▶ Identify the software applications required. At a minimum you'll need energy management and an analytic application, such as fault detection and diagnostics.

▶ Evaluate whether an upgrade is justified. Take into account maintenance cost on the older control system, and the energy savings and potential utility rebates and incentives on the new control system.

The performance level of a building is directly related to the performance level of its control systems. One cannot manage a high performance

Figure 21.2

Figure 21.3

building without high performing control systems. The importance of maintenance of these systems cannot be stressed enough. Regular preventive maintenance assures that the control systems and equipment operate well, extends the equipment lifecycle, minimizes repairs, and affects energy consumption and occupant comfort.

About the Author

Mr. Sinopoli is an innovator in the high performance building industry. For over thirty years, he has designed and engineered operationally efficient, intuitive, and sustainable buildings through an integrated design matrix of building systems and technology. His design work can be found in many building types and uses throughout the United States, Asia, Europe, the Middle East, South America, and Africa. He has consulted and lectured government organizations, industry associations, and Fortune 500 companies.

Mr. Sinopoli has experience in the healthcare, corporate, education, manufacturing, finance, construction, and government industry sectors. His clients have included Fortune 100 corporations, the United States Postal Service, the United States Air Force, major K-12 school districts throughout the country, statewide university systems, airports and ports, the Internal Revenue Service, large private and public hospitals, technology companies, and nationwide developers.

Mr. Sinopoli has experience in over 400 construction projects including design, project management, energy usage, contract administration, and quality assurance with experience in multiple building types, such as high performance building design, energy management systems, and integrated automation. His projects have included the Los Angeles Worldwide Airports (LAWA) conveyance monitoring system; the Ave Maria University building system and IT design; and construction of the Cleveland Clinic Abu Dhabi. He has a B.S. in engineering from Purdue University and a Masters in environmental management from Governor's State University.

Index

Recent Artech House Titles in Power Engineering

Andres Carvallo, Series Editor

Signal Processing for RF Circuit Impairment Mitigation in Wireless Communications, Xinping Huang, Zhiwen Zhu, and Henry Leung

Synergies for Sustainable Energy, Elvin Yüzügüllü

Telecommunication Networks for the Smart Grid, Alberto Sendin, Miguel A. Sanchez-Fornie, Iñigo Berganza, Javier Simon, and Iker Urrutia

For further information on these and other Artech House titles, including previously considered out-of-print books now available through our In-Print-Forever® (IPF®) program, contact:

Artech House	Artech House
685 Canton Street	16 Sussex Street
Norwood, MA 02062	London SW1V 4RW UK
Phone: 781-769-9750	Phone: +44 (0)20 7596-8750
Fax: 781-769-6334	Fax: +44 (0)20 7630-0166
e-mail: artech@artechhouse.com	e-mail: artech-uk@artechhouse.com

Find us on the World Wide Web at: www.artechhouse.com